WHERE THE WILDNESS PLEASES

Where the Wildness Pleases

THE ENGLISH GARDEN CELEBRATED

Caroline Holmes

ACC ART BOOKS

CONTENTS

Rock Walk with foxgloves, Wakehurst Place
The Rock Walk at Wakehurst glimpsed through foxgloves. It is one of the sandrock outcrops that are a distinctive feature of the High Weald, formed 130 million years ago as rivers deposited beds of sand and clay across a vast plain. As layer after layer was deposited South East England rose up to become a broad dome compacting the earliest deposits to form sandstone. In turn the later deposits of sand and clay were eroded away revealing the plateau that is now the High Weald. Under the abundant tree canopy, combined with relatively high rainfall, mosses, liverworts and ferns thrived in the humidity. Sandrock was later valued as Wealden sandstone for building not least at Battle Abbey, Bateman's and Wakehurst Place. (Wakehurst, Jim Holden © RBG Kew)

INTRODUCTION

My Saxon Weald! My cool and candid Weald!
Dear God! the heart, the very heart, of me
That plays and strays a truant in strange lands,
Always returns and finds its inward peace,
Its swing of truth, its measure of restraint,
Here among meadows, orchards, lanes and shaws.

'Night' by Vita Sackville-West

'The English Garden' evokes visions of plants flourishing under a gentle climate, feeding on ancient soils in a well-worked landscape. Where to set the celebrations for an English garden? 'The Garden of England' – Kent – seems apt but I have chosen to climb a little and enter the glorious High Weald, which includes East Sussex, for an overview of what has been designated an Area of Outstanding Natural Beauty since 1983. Its legendary woodlands conceal and reveal dramatic landscapes amongst which castles, mansions and cottages have carved out horticultural settings. It enjoys higher sunshine levels than much of England. Gardens range from grand in design and glorious eccentricity, to arts and crafts and plots of ingenuity – English not in the linguistic sense but with the climatic opportunities to be global in their planting.

Sandwiched between the North and South Downs, the geology of the High Weald ranges from its famous lowland-associated Weald Clay, running alongside Hastings' sand to the coastal cliffs. The growth and use of the English oak was part of the Wealdsman's DNA, its constant presence leading it to be referred to as the Sussex weed. Lying on its margins, the famed gardens of Sissinghurst Castle look up to the High Weald. *Quercus robur* (*robur*, Latin for 'oak-wood and strength') is the specific epithet selected by Linnaeus for the indigenous English oak. Strong in a mineral sense, with iron that lay beneath the soil, its exploitation fuelled finances, building and agriculture, and the exceptional strength of Wealden oak is attributed to the ferruginous clay. The High Weald undulates to a short coastline between Hastings and Winchelsea (witness to the arrival of the Normans who marched west to what is known as Battle). In name three ancient forests survive: Ashdown, Bedgebury and St Leonard's, and three rivers debouch into its short coastline: the Brede, the Rother and the Tillingham. Across the High Weald gardeners can borrow the landscape for miles or enclose their plots with architectural practicality or conceit, shaped and outlined with shrubs and roses. The mighty oak is ubiquitous, punctuating the scene with a delicate and yet determined solidity that suggest dryads may be nearby. A.A. Milne's Hundred Acre Wood is a reality within Ashdown Forest, and it is no surprise to learn that at Bateman's Rudyard Kipling was inspired to write *Puck of Pook's Hill.*

Flower garden to upper lake, Gravetye Manor
William Robinson gardened words, flowers and trees alike with a hardy approach that lives on at his former home, Gravetye Manor. Herbaceous perennials, a lichen-encrusted garden wall with a view to lakes across a flowery meadow and the High Weald beyond encapsulate the elements that make a celebrated English garden.

BATTLE ABBEY TO THE SACKVILLE-WESTS

Prehistoric man and the Romans were there first. Then, on 14 October 1066 the Norsemen arrived near Hastings. At this point William of Normandy and his nobles were ambitious and confident but yet to be all-conquering. Their decisive victory at the Battle of Hastings was followed by dramatic changes in the landowning classes across England, from Anglo-Saxon to Norman. In atonement for the loss of life, Battle Abbey was founded in 1071, established as a Benedictine order with its high altar placed on the spot where King Harold was slain. Dedicated to the Trinity, the Virgin Mary and St Martin of Tours, the daily lives of Benedictine monks included gardening in which they were innovative and successful. In the case of Battle, much produce was sold at local markets. In 1538, following the Dissolution of the Monasteries, the rich and important Abbey was given to Sir Anthony Browne who, in an act of wanton vandalism, demolished the church, converting the guest house and abbot's house into a magnificent residence. In 1786, at the height of popularity of Gothic horror novels, the 'aged debauchee' 49-year-old Godfrey Webster married 15-year-old Elizabeth Vassall, a Jamaican sugar plantation heiress. In 1810 they divorced; he gained an annual income of £7,000 which he used to renovate the abbey, and build a Gothic dairy and icehouse. From 1857 to 1901 the Duke and Duchess of Cleveland

Fairlight End garden

Fairlight End garden lies on the edge of the High Weald with views on a clear day to the coast north-east of Hastings. The control and design of the garden dissolves not only in the morning mist but in the design and plantings. A contemporary approach that would have delighted William Robinson at Gravetye Manor. (Courtesy Robin Hutt)

owned the estate; she laid out elaborate formal gardens – her orchard has been replanted.

The Sackvilles (one of the new Norman and surviving landowners) created the extensive estate of Knole in the Low Weald. Its most famous daughter in the world of gardens, Vita Sackville-West, wished to be remembered as a poet. In May 1930 she and her husband Harold Nicolson bought Sissinghurst Castle (including Castle Farm) on the margins of the High Weald for £12,375. She created a writing room in the tower that dates back to 1565, from where she could observe the architectural control created by her husband, which was softened by her voluptuous plantings. Their play of formality and informality, the notion of garden rooms, white plants by moonlight and kaleidoscopic cottage planting are hallmarks of 20th-century garden design. The family connection continues both in the garden and in prose: her grandson Adam Nicolson is married to the influential gardener and master of all things floral, Sarah Raven. Their home, garden and enterprise, Perch Hill Farm, sits scenically within the High Weald en route to some of its finest follies at Brightling. Adam Nicolson's elegiac *Sissinghurst* captures the joys of landscape and freedom while tracing its history; by contrast, his short book *The Hated Wife*, about Rudyard Kipling's wife Carrie, places the setting of Bateman's in an enigmatic and emotional roller-coaster.

What about all the unchanging centuries on the High Weald? In the 1390s substantial quantities of pedunculate oak were supplied for the vast hammerbeam roof of Westminster Hall. Such was the untouched nature of the region, not least the lack of a single proven Lancelot 'Capability' Brown landscape, that from 1915 to 1923 further supplies were sent from the same source for the restoration of the roof. Damage by German bombers led to additional consignments being despatched in 1941. A pattern of villages with hovels, cottages and yeoman houses surrounded by enormously deep, sheltering hedgerows, small fields and stands of trees, evolved away from mainstream Britain, much to Horace Walpole's disgust. In August 1752 he wrote: '. . . *if you love good roads, conveniences, good inns, plenty of postilions and horses, be so kind as never to go into Sussex. We thought ourselves in the northest part of England; the whole country has a Saxon air, and the inhabitants are as savage* . . .' Iron and its associated industries worked other patterns onto the landscape, but it is not until the 19th century that industrially rich and successful outsiders seek out the unspoilt countryside to build substantial houses and plant exotic tree, shrub and flower collections, assisted by the arrival of railway links.

How perceptions deceive! Little did E.M. Forster know when he wrote in *A Room with a View*: '*Ah, how beautiful the Weald looks! The hills stood out above its radiance, as Fiesole stands above the Tuscan plain . . . She might be forgetting her Italy, but she was noticing more things in her England. One could play a new game with a view, and try to find in its innumerable folds, some town or village that would do for Florence . . .*' Centuries of forest and woodland frame the ancient sylvan Wealden views and folds. The woods that cover the hillsides below Fiesole, framing the view to Florence, were the new plantings by 18th-century British and American residents; when Leonardo da Vinci experimented with flight here it was from barren deforested hillsides.

The Great Storm of 1987 devastated the South East landscape; a toll of the trees lost is a running motif. Today it can be appreciated that nature cleared the canvas for replanting and new designs, and that network comprising stands of trees, thick hedges and small fields is still witnessed throughout the region when travelling on foot, by bicycle, in a car or on a train. Within its framework new ideas have blossomed and taken root.

HEARTS OF OAK OR SUSSEX WEED? HIGH WEALD TREES

Travelling through the High Weald offers an experience rarely enjoyed so close to London – views of wooded hills, framed by gnarled trunks and leafy boughs, and the soft leafy trails that link the indigenous gardens. Many landscapes in this book have used their naturally arboreal settings to offer shelter to global collections of trees. The first landscape-colonisers were the light-seeded birch, aspen and sallow which can withstand cold temperatures; there are pollen deposits dating back 6,000 years for the native lime or linden (*Tilia cordata*), when it was one of the most common flowering trees, but is now effectively extinct. Step by step, new colonisers took root: pine and hazel, alder and oak, ash and elm; the latter devastated by Dutch elm disease in the 1970s. Lastly holly, beech, hornbeam and maple became part of the natural woodland. Bark is an increasingly appreciated aspect of trees – several have connections in this book, such as *Acer davidii* 'George Forrest' a maple with a smooth snakeskin-patterned green bark, laced with silvery-grey, yellow and white markings.

Oak (*Quercus robur*)
The English oak was adopted in leaf form by the National Trust who own/manage seven of the landscapes featured in this book: Sheffield Park has an oak tree dating to 1485. In an exchange of letters in 1938 between Hillier Nurseries and Stephenson Clarke of Borde Hill, the latter asks for an extensive list of *Quercus*; his comments on the letter question whether they came from Oscar Warburg's collection at Boidier near Headley in Surrey (a veritable international gathering of oaks: *Quercus aegilops, Q. humilis, Q. nigra, Q. tomentella, Q. baronii, Q. championii, Q. edithiae, Q. neglecta, Q. stenophylla, Q. suchiensis, Q. thomsoniana, Q. cornea, Q. henryi* and *Q. spicata* var. *brevipetiolata*). Jeremy Bray of High Beeches, where an increasing global collection grows, is a member of the US-based International Oak Society.

The Elizabethan herbalist John Gerard noted that the oak apple was used as a kind of augury by Kentishmen. If you found an ant then that was good news – there would be plenty of grain – but a white worm predicted

Sidney Oak canopy, Penshurst Place
Looking up into the canopy of an ancient oak, one with a legend in the park surrounding Penshurst Place. It is said that this tree grew from one of the acorns sown for the christening of Sir Philip Sidney in 1554. Acorns in profusion fall every year but to ensure the survival of its genes, a small version from a cutting grows in its shadow.

(Left) Wealden oak, *Thuja koraiensis*, *Cunninghamia*, Bedgebury Pinetum
A Wealden oak centre background at Bedgebury Pinetum framed by coniferous pines. On the right the Korean arbor-vitae, *Thuja koraiensis*, and on the left the Formosan Cunninghamia (*Cunninghamia konishii*) introduced by Admiral Clinton-Baker in 1910 from Formosa (present day Taiwan) and planted in 1996.

(Below) Autumn reflections, Marshal's Lake, Bedgebury Pinetum
The blue autumnal skies colour the mirror surface of Marshal's Lake at Bedgebury Pinetum which in turn reflects the mature stands of evergreen and deciduous trees. (Courtesy David Jenner)

(Opposite page, below) *Liquidambar*, Falconhurst
The autumnal fires of the *Liquidambar* are a reminder of its eastern North American origins whilst marking the transition from the horticultural delights of Falconhurst's gardens to the textural delights of the High Weald. Referring to Sheffield Park in his 1934 article, Arthur Soames wrote: '*Liquidambars with their toes in the water colour better here than in a drier position . . . but grow taller in drier ground*'.

the cattle disease murrain, and a spider meant pestilence. Of more practical use, the oak gall was, and still is, used for making ink. Gerard noted that acorns were a good antidote to poisons. Oh, that the oak could protect itself against the oak jewel beetle (*Agrilus biguttatus*), whose females lay eggs within the cracks in the bark! The hatched larvae bore into the trunk, harming the tree's feeding system to the point where the tree cannot survive. The symptoms are black fluid weeping from cracks in the trunk created by bacteria, resulting in an acute decline affecting thousands of trees.

Leonardslee from the air
An aerial view of Leonardslee where the microclimates and topography have supported a world-class collection of trees and shrubs. A microcosm of the High Weald's horticultural development from the 19th century onwards. (Courtesy Leonardslee Garden)

Pine (*Pinus spp.*)

> *There stand these rich men's houses among the pine trees and gardens, and not even the pine trees and gardens can make them tolerable.*
> *Art, Wealth, and Riches* by William Morris

The canopy of the Scots pine (*Pinus sylvestris*) viewed from below is a magnificent sight; its characteristic trunk textured in furrows and platelets like a dry riverbed, becoming redder as the branches reach up to the sky. This and a large Monterey pine (*Pinus radiata*) have provided a dramatic borrowed landscape for Balmoral Cottage.

On his return from a three-year journey (1768–71) aboard HMS *Endeavour* with Captain James Cook, Sir Joseph Banks was appointed the first (unpaid) director of the Royal Botanic Gardens, Kew and was later a guest at Sheffield Park. This voyage was preceded by one to Newfoundland in 1766, where he explored its northern boreal forest, describing the Jack pine that would be named for him – *Pinus banksiana*. Although it can grow to 23m there is a good selection of smaller forms rejoicing in names such as 'Tucker's Dwarf' and 'Uncle Fogy'. The black or Corsican pine (*Pinus nigra*) thrives in the shelter of established woodland gardens such as Borde Hill, where it was introduced in 1890. Frank Kingdon-Ward plant-hunted for the Stephenson Clarkes of Borde Hill. In their archives there is a list of over 60 batches of numbered seeds that he brought back from his 1937–38 Yü expedition, including three separate batches of *P. yunnanensis* seeds.

Using pine as a common name widens the scope to the Australasian Norfolk Island pine (*Araucaria heterophylla*), its straight trunk prized for ship's masts, and Wollemi pines (*Wollemia nobilis*). Both are related to the South American monkey puzzle tree (*Araucaria araucana*).

***Fagus sylvatica* 'Aspleniifolia', High Beeches**

A cut-leafed beech (*Fagus sylvatica* 'Aspleniifolia') at High Beeches in the autumn, marking the route from the central pond as it climbs towards the Magnolia Garden. The light on the golden copper leaves enhanced by their narrow ferny foliage.

Beech (*Fagus*)

Britain's native beech (*Fagus sylvatica*) is a large, noble tree which thrives on calcareous soils, its smooth grey trunk supporting a leafy canopy whose foliage turns a rich copper-gold in the autumn. There are many named cultivars, such as the cut-leafed beech (*Fagus sylvatica* 'Aspleniifolia') and a tricolour European beech (*Fagus sylvatica* 'Purpurea Tricolor'), at Leonardslee. The southern beeches (*Nothofagus*) were discovered by Joseph Banks after he bullied James Cook into stopping off at Tierra del Fuego in 1769. In four hours he collected more than 100 plants, including what he named *Fagus antarctica* (now *Nothofagus antarctica)*, the Chilean fire bush. Wakehurst holds the national collection.

Japanese maple (*Acer*)

The colours and forms of Japanese maples – principally forms of *A. japonicum* and *A. palmatum* – received a popular welcome after their first introduction in 1864 and 1820, pleasing the plant-collector landowners (see pages 165–169) as well as Margaret Beale at arts and crafts-inspired Standen. William Robinson at Gravetye Manor preferred the hardier, established Norway and Virginian maples, *A. platanoides* and *A. saccharum*. When Frank Kingdon-Ward plant-hunted for the Stephenson Clarkes at Borde Hill, amongst the numbered seeds he brought back from his 1937–38 Chinese Yü expedition were *Acer cappadocicum* var. *sinicum, A. erosum, A. giraldii* and *A. tetramerum* var. *betulifolium*, all with notable foliage.

***Acer palmatum* 'Dissectum', purple, sculptural, Borde Hill**

A venerable exquisitely shaped *Acer palmatum* 'Dissectum' at Borde Hill, centred in the Long Border stepping down from the Italian Garden. Its gnarled, twisted trunk and pruning echo its Japanese roots. It is prized by gardeners for its fine, lace-like filigree leaves here tinged with pink in May.

Woodland Garden, bluebells, Hole Park
The Woodland Garden at Hole Park appears effortlessly naturalised here in April with a thick carpet of bluebells amongst the coppiced trees.

***Acer shirasawanum* and *Acer palmatum*, High Beeches**
Acer shirasawanum forms a moonlight-yellow haze behind an *Acer palmatum* in the low autumnal sunshine at High Beeches.

CHAPTER ONE

Keeping it in the family

Barham coat of arms, bear in pebbles, Hole Park
Looking as though he has stepped out of a picture book, the muzzled bear depicted in pebbles represents the Barham family coat of arms at Hole Park.

Two ancient and noble houses and gardens lie just six miles apart in Wealden Kent: the Low Weald shelters the Sackvilles' Knole, surrounded by a vast deer park, in Sevenoaks, while on the High Weald Penshurst dominates – palatial in concept and described by John Julius Norwich as '*the grandest and most perfectly preserved example of an unfortified manor house in all England*.' The 19th century witnessed the 'discovery' of the High Weald for its stunning topography and interesting microclimates, enticing successful Victorian families to put down root. The descendants of four still flourish: two from great plant hunters and collectors – the Stephenson Clarkes at Borde Hill, and the Boscawens and Brays at High Beeches. Two continue the legacy of family home and successful estate – the Barhams at Hole Park and the Talbots at Falconhurst.

Loder is a family name that shaped three celebrated gardens – High Beeches, Leonardslee and Wakehurst Place. Born in St Petersburg in 1823, on the death of his father in 1871, Robert Loder inherited extensive estates in Sussex (where he bought High Beeches) and Northamptonshire (he later combined both counties by becoming 1st Baronet of Whittlebury and High Beeches), as well as estates in Russia and Sweden. He had 10 children, two of whom were to become famed horticulturists and plant collectors: Sir Edmund Loder, who married Marion Hubbard of Leonardslee, and Gerald Loder, later 1st Lord Wakehurst. Gerald Loder established collections of rare plants, especially rhododendrons, introduced from China, North Burma (now Myanmar) and Tibet, at Wakehurst as well as generously supporting Kew when it was creating the first National Pinetum at Bedgebury in the 1920s. His unique collection of plants from New Zealand and Australasia inspired him to present a cup to be awarded annually in New Zealand for the best collection of native plants. A Fellow of the Linnean Society from 1914, he served on its council before he became Vice President. In 1921 he presented the Loder Cup to the Royal Horticultural Society to commemorate his elder brother, Sir Edmund Loder's exemplary work at Leonardslee, successfully growing hybrid rhododendrons and conifers. From 1926 to 1927 he was President of the Royal English Arboricultural Society, was closely involved in the 1931 Conifer Conference, and in 1936 he was awarded the RHS Victoria Medal of Honour. The family name is horticulturally memorialised in the Rhododendron Loderi Group resulting from the hybridisation of Robert Fortune's Himalayan *R. griffithianum* with the Chinese *R. auriculatum*.

Azalea ring, *Rhododendron luteum*, Borde Hill
As well as being a beacon of spring colour the yellow-flowered *Rhododendron luteum* has a wonderful wafting scent. Here among the floriferous Azalea Ring at Borde Hill it is making its presence felt in contrasting, zinging colour against the muted whites and pinks. (Courtesy Borde Hill)

PENSHURST PLACE

Now, Penshurst, they that will proportion thee
With other edifices, when they see
Those proud, ambitious heaps, and nothing else,
May say their lords have built, but thy lord dwells.
To Penshurst by Ben Jonson

Union Jack garden, mount and heraldic parterre Penshurst'
The mount was a favourite conceit in Tudor and Jacobean gardens, providing a raised platform from which to admire the formal layout plants. At Penshurst this makes the perfect vantage point to view the large red, white and blue flowery parterre inspired by the Union Jack.

Droveways are the most ancient of routes through the High Weald. Adopted and improved by the Romans and actively used by the Saxons and Jutes, churches sprung up along their course, especially by river crossings, which in turn dictated the founding of villages such as Penshurst. Of the range of building stones quarried on the High Weald, both Penshurst Place and Wakehurst Place were built in 'country stone', a golden sandstone streaked with ruddy veining. In 1338 John de Pulteney, draper and four times Mayor of London, acquired the estate of Penshurst. The roof of his earliest building (de Pulteney's now Baron's Hall) was built with another Wealden tree – the sweet chestnut (*Castanea sativa*). Building near a main road was viewed as an advantage because travellers could be stopped for the latest news.

Six hundred years later, the Great Storm of 1987 destroyed thousands of trees on the Penshurst estate; 10,000 were planted as replacements. In 2012 the Diamond Jubilee of Her Majesty the Queen was marked by the restoration of a 16th-century plantation of 60 sweet chestnuts in a diamond-shaped copse surrounding four saplings from the Sidney Oak. The traditional symbol of peace carried by Kentish men dating back to the arrival of the Normans, the holm oak (*Quercus ilex*), is planted at each point of the diamond.

> *Though she was very English – and made great play with that – she was a woman of the Renaissance . . . We were, in the Elizabethan Age, a small people of only some five millions. But a Society that had Elizabeth at its head, Burghley as its statesman, Philip Sidney as its pattern of chivalry, Drake for its sea captain; whose poet was Spenser, whose philosopher Bacon; that had Shakespeare for its dramatist, may well bear comparison, so far as civilisation is concerned, with any of the larger societies of our time.*[1]

Penshurst was home to Sir Philip Sidney; it is said that acorns were planted at his christening in 1554. He was the third generation, and 13 generations later the family continues with Philip, 2nd Viscount De L'Isle; his father William planted acorns from the same tree around the world. Due to its age, cuttings have been taken so that its clone can survive not least in the Queen's Diamond Jubilee plantation. Records show that de Pulteney had

1. A.L. Rowse, *The England of Elizabeth* (London: Macmillan, 1950).

Italian Garden, pool, statue of Hercules, Penshurst Place
Created in the 1560s, Penshurst's pool in the Italian Garden is a clever piece of Renaissance illusion – it appears to be circular when viewed from the house's piano nobile. A statue of Hercules looks out.

established a garden by 1346. Two hundred years later Tudor Renaissance styles and fashions were adopted with gusto by Philip Sidney's father Henry and mother Mary (formerly Lady Mary Dudley, sister of the great Robert Dudley beloved of Queen Elizabeth I and creator of the magnificent Kenilworth Castle gardens). In the reverse of the experience of today's visitor, guests would be invited into the house's *piano nobile* to admire the high formality of the Italian Garden created in the 1560s, brilliantly conceived as a trompe l'oeil so that it appears circular. Stepping from the house a still-extant network of walls and terraces was created from which, amongst many, Gloriana – Queen Elizabeth I, enjoyed sensational views. Renaissance gardening coined the phrase '*Nature produces better fruit if planted and cultivated*'. Sir Robert Sidney (who became the Earl of Leicester) completed his father's orchard walls for a range of improved fruits such as varieties of apricots, peaches, plums and apples. Ben Jonson's *On Penshurst*, written in the tradition of country house poetry, oozes the fecundity and fun encapsulated in the gardens and their produce.

In keeping with much of the High Weald, from the mid-18th century Penshurst, its terraces and multibank divisions slumbered for nearly 100 years to be awoken by a new young master, Philip Sidney, 1st Baron De L'Isle and Dudley, known as Lord De L'Isle and Dudley, whose status was considerably enhanced by his marriage to Lady Sophia FitzClarence, daughter of King William IV and the actress Mrs Dorothea Jordan. At a time when Victorian Italianate styles were at their zenith under

The Porcupine by Robert Rattray, Penshurst Place
In his unmistakeable style, Robert Rattray fashioned this bristling ironwork into the Sidney family emblem of the porcupine to celebrate the millennium. The house and Heraldic Garden are liberally decorated with heraldic symbols relating to family members and their marriages. Elsewhere a topiary bear with a ragged staff, the Dudley heraldic animal, flanks Rattray's 'The Archer' appropriately wielding a broad arrow from the Sidney coat of arms. Look out for the Gamage Welsh griffin, the Gauntlet for the Perry family, Falcon for the Duke of Bedford who formerly owned Penshurst, the House of Beaufort portcullis and the Earls of Warwick's rampant double-tailed lion.

Magnolia Garden, Penshurst Place
The Magnolia Garden is dedicated to the memory of Jacqueline, Viscountess De L'Isle (1914–1962). Lord De L'Isle gave the original Naiad sculpture to Government House in honour of the Bicentenary of Australia in February 1988. This bronze version by Patience Devas with the inscription on the marble surround '*MUTUIS ANIMIS AMANTE AMANTUR*' is taken from the Latin poet Catullus ('. . . with mutual passions they love and are loved . . .') from which Sir Philip Sidney's verse beginning '*My true love hath my heart and I have his*' may have been derived. The original statue overlooks Lake Burley Griffin, from the grounds of Government House, Canberra where Lady De L'Isle died in November 1962 while her husband was Governor-General. (Courtesy Penshurst)

Heraldic Garden, Penshurst Place
Lining the view from the gardens to the house with a contemporary echo of Tudor gardens, there are heraldic beasts on chequered posts. The Sidney symbols include the bear with the ragged staff and the porcupine, which also appear as topiaries in the gardens.

Charles Barry and W.A. Nesfield, the Italian Garden was reborn: starting with the Renaissance lily pond a series of garden rooms were revealed and concealed by close-clipped yew hedges. Victorian tastes were further fulfilled by a magnificent kitchen garden. In the 1940s Virginia Woolf and Vita Sackville-West visited, the former saying that '*all the seeds of English fiction lie latent*' in Philip Sidney's *Arcadia*.[2]

The latest instalment starts in 1945 after the vicissitudes of World War II. The Italian Garden had survived and was revived whilst the wider gardens benefitted from a range of floral, seasonal and design ideas. In good tradition, Penshurst's weathered façades are flanked by the family's private gardens on three sides. There are clearly marked footpaths and bridleways across the estate. Today's visitor approaches from the east. Garden enthusiasts should enter the walled garden to make a floral approach through a series of gardens from which is glimpsed the house. The red, white and blue-themed Union Flag Garden follows, around which are displayed heraldic topiary and Robert Rattray's Archer sun-dial and Porcupine statue. George Carter's double herbaceous borders triumphantly line the route to the Italian Garden, from which steps lead into the Garden Tower and on to Baron's Hall. Inspired by the medieval, mount the staircase and enjoy a Renaissance view from the *piano nobile*.

2. Virginia Woolf, *The Common Reader*, Second Series (London: Hogarth Press, 1959), p. 49.

BORDE HILL

'. . . as fair a flower garden as we could desire . . .'
Country Life, 1902

The earliest record of ownership is for Stephen Borde, dating from 1534. He had land in 29 parishes with 19 manors; he died in 1557. His brother, the original 'Merry Andrew', was Henry VIII's court jester and physician; amongst much advice on herbal medicine he is attributed with introducing hops into brewing. Like Penshurst Place, Borde Hill House (built between 1598 and 1600) is one of the few historic grand houses on the High Weald. A suitable domain for another Stephen Borde, knighted in 1603, it is referred to as Borde Hill for the first time on the christening of his daughter in 1606. In 1720 William Borde died without an heir, and the estate was split and sold.

In 1870 a sale catalogue and map were prepared

Emmenopterys henryi, flowers, Borde Hill
The very rare *Emmenopterys henryi* was introduced from China by Ernest Wilson in 1907 and described as: '*one of the most strikingly beautiful trees of the Chinese forest, with its flattish to pyramidal corymbs of white, rather large flowers and still larger white bracts.*' The Greek origin of its name *ĕmmĕnēs* enduring and *ptĕryx* ('a wing') reflects the large and winged lobes of the calyx. Augustine Henry preceded Wilson, as a plant collector and dendrologist, through central China and Formosa (modern day Taiwan). It is so exceedingly shy about producing its creamy white fragrant flowers that Borde Hill had to wait for its first spectacular appearance until 2011. It was repeated in 2018. (Courtesy Borde Hill)

showing an affluent property. The house had recently been extended in surroundings of ornamental pleasure grounds with a croquet lawn, parterre, shrubberies, evergreens and ornamental timber. In 1890 new tree planting included Corsican pines (*Pinus nigra*) and additional oak woodlands. Three years later, Mrs Cunliffe Lister sold Borde Hill House with 676 acres to Colonel Stephenson Robert Clarke – Stephie to his friends. The Newcastle-based firm of Stephenson Clarke worked in coal mining and shipping, later freight rail, reputedly running the largest collier fleet and the greatest number of coal rail trucks in Britain. After the nationalisation of the railways the company diversified into, among other ventures, the hansom-cab business in Tooting. As increasingly successful entrepreneurs, they moved first to Croydon Lodge in Tooting, then to Brook House in Sussex. Following his marriage to Edith Gertrude Godman they moved into Borde Hill House.

Borde Hill sits on three ridges that run roughly east–west, with soil types ranging from a Wealden clay substrata and Tunbridge Wells sand, to rich loams. Like so many of the plantsman's gardens in the vicinity, its topography leads to variable exposure and a wide range of microclimates. Careful observation enabled the successful growing of a global range of trees and shrubs, especially those from China and South America. The Tree

Midsummer Border, Borde Hill

The Midsummer Border, designed by Dr Tony Lord, is a vibrant mixture of the red and yellow perennials *Crocosmia* and *Hemerocallis*, the drama of the deep purple and pink of the sedums adding texture. Waves of green and gold are added by the grasses *Calamagrostis brachytricha* and *Miscanthus sinensis* 'Gracillimus'. (Courtesy Borde Hill)

Register of the British Isles is a charitable organisation that publishes details of rare, historic, champion and other significant trees. In 1999, Dr Owen Johnson wrote:

> *The Borde Hill estate contains more Champion Trees – the tallest or largest of their kind recorded in the British Isles – than any other privately-owned collection in the country. Many of these species are exceedingly rare, both in cultivation and in their natural habitats. Others are examples of common trees, which have thrived in the parkland conditions at Borde Hill. The collection is of great horticultural and conservation importance, and also constitutes one of the finest historical garden landscapes in the country.*

'*The greatest amateur all-rounder in the gardening world of the twentieth century*' was how Harold Hillier described Stephenson Robert Clarke, who had an excellent relationship with the Hillier, Veitch and Waterer nurseries. The Borde Hill estate was a primed canvas for an already enthusiastic collector and keen naturalist. He was a significant sponsor of plant hunters (see pages 165–169), raising many of their horticultural discoveries. An avid correspondent and horticultural networker, there were active exchanges between Borde Hill and neighbouring Nymans and Leonardslee, as well as with the famed rhododendron gardens at Exbury created by Lionel de Rothschild. Many animated letters survive from his good friend John Charles Williams (1861–1939) of Caerhays, who devoted his life to the cultivation and study of Chinese flora, especially rhododendrons, magnolias and camellias. In 1931 Williams sent Stephenson Robert Clarke a *Magnolia sprengeri* which had been raised from seed brought back by 'Chinese' Wilson in 1908. The beauty of its rich pink flowers inspired its common name of 'goddess magnolia', and *M. sprengeri* var. *diva* is still a showstopper in March and April. The hugely popular *Camellia* x *williamsii* 'Donation', which was first exhibited in 1936 and won the RHS Award of Garden Merit in 1941, was raised by Stephenson Robert Clarke when he crossed *C. japonica* 'Donckelarii' with *C. saluenensis*.

Stephenson Robert Clarke created a ha-ha and the South Lawn, thus opening up gentle views that contrast in a very 19th-century way with those to the north, through and over dramatic woodland to the Balcombe railway viaduct. The East Garden was distinctly floral and colourful with roses, lilies, hollyhocks, veronica, red lychnis, dark-blue monkshood, phlox, peonies, larkspur and foxgloves. The Rock Garden he created was noted for its Japanese water iris (*Iris ensata*). From 1905 Warren Wood hosted many of the new species sent by plant hunters, as did the Roadside Plantation alongside Borde Hill Lane. In 1910 plantings extended into Little Bentley Wood and Stephanie's Glade. In 1906 that stalwart of Edwardian horticultural excellence, a walled garden was created for vegetable production, and the house was further extended in 1912.

Borde Hill's archives have many lists relating to Stephenson Robert Clarke's sponsorship of plant hunters. In 1925 Frank Kingdon-Ward's rhododendrons were planted in the newly formed Garden of Allah in the North Park gardens. In the 1920s George Forrest sent seeds of one of the first Asiatic magnolia species, *M. campbellii*, which graces Borde Hill at a height of over 19 metres. Between 1927 and 1937 The Tolls and The Gorse were planted up, the gardens were extended to the north and south and finally Gore's Wood was planted. In 1932, following the death of Vicary Gibbs of Aldenham House, Radlett, Hertfordshire, there was a sale of his rare shrubs and trees from which Stephenson Robert Clarke bought many rarities including the uncommon summer-flowering *Meliosma beaniana*. A jaw-dropping record of Stephenson Robert Clarke's achievements was published by Oxford University Press in 1935: *Catalogue of the Trees and Shrubs (excluding Rhododendrons) at Borde Hill, Sussex, in December 1932*, compiled by Albert Bruce Jackson, A.L.S., Veitch Memorial Medallist (1923), with foreword and notes by Colonel Stephenson Robert Clarke. It runs to 275 pages, starting with *Abelia engleriana* and including the familiar and unfamiliar, such as *Betula*, *Carya*, *Chamaecyparis*, *Cupressus*, *Juniperus*, *Larix*, *Magnolia*, *Malus*, *Picea*, *Pinus*, *Pterocarya*, *Quercus*, *Sorbus*, *Tilia*, *Tsuga*, closing with *Zenobia speciosa*. The 'excluding Rhododendrons' accounts for a further 1,000 varieties.

In 1948 Stephenson Robert Clarke was succeeded by his son Colonel Sir Ralph Stephenson Clarke who, with his wife Rebecca (Becky), revived the gardens and landscape following World War II. By 1965 it had been decided to incorporate the garden as a registered charity to ensure its survival, and to open it to the public. In 1968 Vita Sackville-West's former gardener Jack Vass was appointed Head Gardener; his contribution until 1979 is marked by the grassy Jack Vass Walk in the Old Rhododendron Garden.

In 1970 Robert Stephenson Clarke inherited, and he continued to add to the tree and rhododendron

Round Dell, Borde Hill

Just as Borde Hill resonates with the history of people, plants and places it is also a setting for contemporary ideas; the garden designer Sophie Walker has been inspired by modern plant hunters. When she designed 'Cave Pavilion' in 2014 she was the youngest woman ever to design a garden for the Chelsea Flower Show. This and 'Valley Garden' at the Hampton Court Palace Flower Show reflected how the wilderness fascinates and attracts the quest to explore nature's green diversity. In contrast her design for the Round Dell counterposes nature with what it cannot do – be linear and directive through a straight line. (Courtesy Borde Hill)

(Following pages) Rose Garden, Borde Hill

The Jay Robins Rose Garden, named for Andrewjohn and Eleni Stephenson Clarke's daughter, was designed in 1996 by Robin Williams. Its English formal setting of box and yew topiary, lavender and *Nepeta* hedges celebrates the glorious scented array of 750 David Austin English Roses. (Courtesy Borde Hill)

collections. His cousin Desmond Clarke was chief editor for the eighth revised edition of W.J. Bean's *Trees & Shrubs Hardy in the British Isles*[3] – unsurprisingly there is frequent mention of Borde Hill's collections. Drought hit during the summers of 1975 and 1976; many of the rhododendrons died, especially smaller species and historic bushes such as Forrest's introduced *R. preptum* (*preptus* rather aptly translates from Greek as 'distinguished or eminent') from Upper Burma (Myanmar) and Yunnan, and *R. heliolepis* from West China and North Burma (Myanmar). First, Michael Zander, then Michael Lear tagged the trees to create a Woody Plant Catalogue, while John McGuire worked on the rhododendron collection from 1980. In 1982 the rectangular Bride's Pool and the West Bank terrace were created.

The Great Storm of October 1987 created havoc. Cuttings were taken from many species that had blown down or were damaged. Borde Hill lost at least 13 coniferous and six broad-leaved champion trees. Of the 2,678 specimen trees, 220 were uprooted or severely damaged. Notable losses were Forrest's *Picea likiangensis* from The Tolls, the rare Chinese *Meliosma pinnata* var. *oldhamii* (introduced in 1900) from Warren Wood, and a champion Algerian fir (*Abies numidica*) in the North Park. On the death of his father, within a month of the catastrophe, Andrewjohn Stephenson Clarke inherited Borde Hill with his wife Eleni. The Robertsmere Lake was created in his father's memory, and one of their first tasks was to update the Woody Plant Catalogue.

In 1992 Borde Hill received an English Heritage grant for replanting following the 1987 devastation. In 1997 a Heritage Lottery Fund grant was obtained to restore and replant the gardens, improve path access, restore the Victorian glasshouses and purchase machinery. The 20th century closed with the South East England Tourist Board Visitor Attraction of the Year Award, and the 21st saw the Sussex Heritage Trust Award for the Victorian greenhouses. The botanical and historical importance of Borde Hill lies in the fact that many of the trees and shrubs growing there are clones or first-generation seedlings raised in this country, thus providing a living record of their provenance from, say, China or Japan.

Borde Hill encapsulates the make or break of so many 19th-century gardens: the zenith of plant collecting under Stephenson Robert Clarke; the dedication of the subsequent two generations; and now, under Andrewjohn and Eleni Stephenson Clarke, new ground is being broken with their commissioning of several contemporary garden designers whose works stand alone as inspiration to the domestic garden whilst weaving different rhythms into the established canvas.

3. W.J. Bean, *Trees & Shrubs Hardy in the British Isles* (London: John Murray, 1989).

HIGH BEECHES

Gardens should be restful to the eye, and full of beauty at all seasons
John G. Millais

Blood and bonemeal are traditional fertilisers for the soil. According to legend, this lies at the root of the great drifts of the sweetly scented lily of the valley (*Convallaria majalis*) that carpet many parts of the High Beeches landscape. Situated on the easterly edges of St Leonard's Forest, it was said to be the domain of the last dragon in England in the sixth century, killed by the appropriately named St Leonard the Dragon Slayer. During the battle, the blood lost by St Leonard seeped into the ground from which sprung lilies of the valley. Today 27 acres of the landscape can be visited; beyond lies the 100-acre Dencombe Wood and a seven-acre wildflower meadow providing a colourful tapestry below the house.

At High Beeches the sense of keeping it in the family is less linear. As detailed above (see page 19), Sir Robert Loder who bought the estate in 1849, was sire to Sir Edmund Loder of Leonardslee and Gerald Loder, 1st Lord Wakehurst of Wakehurst Place. Sir Robert's second son Wilfred inherited, followed in 1906 by Wilfred's 22-year-old son (later Colonel) Giles Loder whose exemplary horticultural pursuits continued until his death in 1966. Edward and Anne Boscawen then bought the property which is now run by their descendants, Jeremy and Sarah Bray. Family in the sense of the nearby Loder cousins, and continuity in the hands of Eric Stockton (who was gardener from 1927 until his death in 1979) and the

***Acer palmatum*, natural seedling, High Beeches**
This Japanese maple (*Acer palmatum*) is a natural seedling not a named cultivar. It was first introduced to the UK in the 1820s. The low sunshine gives a painterly quality. Just the place to sit and admire your handiwork, the Brays encourage visitors to come to High Beeches for peace and tranquillity.

Rhododendron eximium, High Beeches
The long, bell-shaped creamy flowers of *Rhododendron eximium*, which forms a large shrub. The young growths are covered with an orange-brown indumentum as are the undersides of the large leaves. Forrest Bank, named for the plant hunter George Forrest, forms a boundary planted with special rhododendrons that have pink or white flowers. (Courtesy High Beeches)

Rhododendron kesangiae, High Beeches
Recently introduced from Bhutan and still rare, *Rhododendron kesangiae* has large trusses of rose fading to pale pink flowers, the indumentum is silvery to fawn. The leaves can be up to 30cm long. At High Beeches, it has been planted near the excellent bell-shape flowered great Chinese rhododendron (*R. sinogrande*). (Courtesy High Beeches)

non-profitmaking High Beeches Gardens Conservation Trust established in 1978 by the Boscawen/Bray families. The Great Storm of 1987 took out 250 trees, including rare specimens. Arboreal descendants arose from the disaster, such as seedlings from a *Tetracentron sinense* that had been raised from one of the first seeds brought back by 'Chinese' Wilson in 1904.

The opening quote was made by John G. Millais, son of the Pre-Raphaelite artist. He was a neighbour of Colonel Giles Loder. His advice on how to produce a harmonious balance between landscape and appreciation of individual plants included not to overcrowd, avoid over concentration on one genus, and use only the best plants. Loder worked against a backdrop of native oaks which the Brays are looking into replacing with trees that are more disease and climate-change resistant. High Beeches enjoys exceptional autumn colour thanks to Boscawen plantings from the late 1980s: *Nyssa sylvatica* (including its cultivar *N. s.* 'High Beeches'), *Liquidambar,* oaks and *Acers* (badly affected by the 2018 drought) for their foliage; for their decorative fruit: *Malus* and *Crataegeus*; and *Kalopanax pictus* for its flowers, which run from late summer into autumn. The formula for August is a cool backdrop of *Eucryphia*, *Stewartia* and *Hydrangea*, with the colourful foliage of *Acers* and the smoke bush *Cotinus* 'Grace', plus the flowers of *Gentiana asclepiadea* and *Lobelia tupa*; the scent of *Clethra* wafts across the scene. This is the only place in the UK where the gentians self-seed.

Named for the plant hunter George Forrest (see pp), who was responsible for the introduction of over 300 new species of rhododendron, Forrest Bank forms a boundary planted with special rhododendrons that flourish pink or white flowers. *Rhododendron falconeri* subsp. *eximium* forms a large shrub; the young growth is covered with an orange-brown indumentum as are the undersides of the large leaves, and the long bell-shaped flowers are cream. Recently introduced from Bhutan is what Sarah describes as a showstopper – *R. kesangiae*. Producing trusses of rose fading to pale pink flowers, with silvery to fawn indumentum, it is planted near the excellent flowering *R. sinogrande*. In addition, there is the smaller, twiggy and aromatic *R. primuliflorum* whose daphne-like pink flowers have a yellow tube, and also the small shrubby *R. racemosum* with pale to bright pink funnel-shaped flowers.

The Brays and their staff of two are actively involved in maintaining High Beeches. All the trees are clearly labelled, and to protect self-set seedlings visitors are asked not to stray from the paths. Motorised 'trucks' are not used, all tending is done using wheelbarrows to prevent compaction. The result is that the paths are gentle indicators unlike those in other parks open to the public. The Brays take a holistic approach to buying new plants, not just collecting rarities for the sake of it, thus ensuring a pleasing balanced canvas.

From the 1970s when Anne Boscawen entered

Colonel's Seat, autumn colour, High Beeches

This is the view from the Colonel's Seat, named for Giles Loder. The autumn colours are mellow in the brightest sense, the moss indicating how untouched the landscape is and how clean the air. Colonel Loder's careful selection of *Rhododendron* still provides the backbone for spring, celebrated in the Loderi Walk. The namesake *R.* 'Loderi' and several of its hybrids are maintained in layers. He was good friends with Arthur Soames at Sheffield Park with whom he shared plant collections.

***Ilex perado*, High Beeches**
The holly *Ilex perado* is very rare in the UK. It has large, evergreen, magnolia-like leaves with clusters of red-orange berries. Tender when young it can grow to 20m in Southern England.

autumn foliage at an RHS show and was awarded a First Class Certificate for *Nyssa sylvatica* 'High Beeches', they have been regular winners. In 1988 *Quercus oxyodon* (from the Eastern Himalayas), one of only two mature specimens in the UK, with an uncharacteristic sweet chestnut-like evergreen foliage and tiny acorns, was awarded the RHS Award of Garden Merit. The acorns take two years to mature but are not viable in the British climate. The flowering dogwood *Cornus* 'Norman Hadden' received an RHS First Class Certificate for an exhibition vase in July 2002. The haul for 2019 was three lots of silverware. Award winners or not, amongst many more, these plants are highly prized: the handsome-leafed *R. macabeanum* with large trusses of bell-shaped, pale yellow flowers blotched with purple, was introduced by Kingdon-Ward *c.*1928 from Manipur. Only a handful of *Magnolia sargentiana* var. *robusta* Chenault form are in existence; the specimen here fell in the 1987 storm but has recovered and produces its rose-crimson 23-centimetre flowers magnificently. Here also are two South Americans: one the sole example of its species, Prince Albert's yew (*Saxegothaea conspicua*), and the Chilean plum-fruited yew (*Prumnopitys andina*).

There is something about the tranquillity at High Beeches that makes you feel you are imbibing a gloriously natural landscape, where you breathe deeply and then find yourself drawn into the wonder of knowledgeable and beautifying planting on an ancient forest ridge. High Beeches is listed Grade II* by English Heritage as an outstanding example of an early 20th-century woodland garden. It holds the national collection of Stewartias. They are in the tea family and closely related to *Camellia* so favour lime-free soil. Their white to cream flowers are produced in succession during July and August, the foliage takes on rich autumnal tints, and the mature trunks are aesthetically pleasing. Only a few of the plants growing at High Beeches have been discussed; the good news is that every plant is labelled and recorded on their website. High Beeches is an RHS Partner Garden and member of the Historic Houses Association.

HOLE PARK

still the cream of the crop

There is a glorious sense of space at Hole Park, with room for the parkland trees (including a Lucombe oak, *Quercus* x *hispanica* 'Lucombeana') to have formed magnificent specimens, framing views that stretch out across the countryside. The house sits elegantly on the drive, with excellent views towards Rolvenden embowered with the Millennium Garden and Tropical Border. Passing the Stable Yard, the Walled Garden provides a delightful stepping-stone into the landscape beyond.

The gardens curve in a large hook: progressing north you pass through the Policy to The Dell and on to the Woodland Garden with its walks and icehouse. A gentle climb to the south brings you back into a yew-hedged green allée, with glorious views to the woodland through the Sundial Garden and up to the Four Seasons Garden and its ponds. The view is released in both directions out across the countryside or over the expansive lawn with the Lucombe oak to the house. The Eagle Slayer statue directs you into a series of typical garden rooms with long borders until you reach W.D. Caröe's Vineyard.

Henry de Hole gave his name to the site in the middle of the 13th century. The Gybbons, residents for nearly 350 years, built a yeoman timber-frame farmhouse in the 16th century, which was later encased in brick. The local wool trade benefitted the family sufficiently that in 1720 they substantially extended the house in Queen Anne style. About 20 years later an icehouse was constructed using the nearby spring-fed stream which flows through the Woodland Walk. It has been restored by the current owners whose own ancestor made his fortune rather appropriately in dairy refrigeration. A hundred years later Thomas Gybbon-Monypenny demolished the original farmhouse, but retained the Queen Anne façade as a central bay for his Elizabethan revival house. Unfortunately, this stretched his finances to bankruptcy and Hole Park was mortgaged. He died a pauper in 1854.

Decades later, the Barham family arrived when Colonel Arthur Barham purchased most of the estate in 1911. Milk and his father George's enterprise had created his fortune: George Barham had seen the opportunity

Carved bears, Hole Park
Although some distance from Ashdown Forest and Winnie-the-Pooh's Hundred Acre Wood, this sculpture of a family of three bears scaling a tree trunk for honey is a delight. It surprises visitors on the Pine Walk in the Woodland Garden at Hole Park. (Courtesy Hole Park)

Memorial gates, Hole Park

Appropriately sombre in tone, these gates designed by the architect W.D. Caroe, take the visitor from the lawns into the walled garden. In 1914 Wilfrid Barham joined the Buffs (the East Kent Regiment), the following year on 9 October he was hit by shrapnel at Ypres and died the next day. Below the dragon and within the ironwork all this is simply recorded in 'WSB, YPRES, OCT10 1915'. How appropriate to look at a Lutyens bench as it was he who played a major part in creating the cemeteries and memorials for the (Imperial) War Graves Commission, working with amongst others Rudyard Kipling of Bateman's who lost his son John. (Courtesy Hole Park)

Eagle Slayer, Hole Park

The Eagle Slayer sculpted by John Bell and hollow cast in bronze by the Coalbrookdale Company of Ironbridge, is a masterpiece of movement and intent. It is an interpretation of one of Aesop's fables where an eagle has just killed a lamb, but the shepherd has retained one of its feathers from an earlier slaughter. Fashioning the feather into an arrow he harnesses his entire frame into powering the bow so that the arrow will soar to its target. In the words of Aesop '*unwittingly we give to our adversaries the means of our own destruction*'. It was originally commissioned for Prince Albert's brainchild The Great Exhibition in 1851. Based in Hyde Park, London for six months it is better known as The Crystal Palace. Six million people marvelled at Britain's technical and industrial might. (Courtesy Hole Park)

to bring much-needed fresh milk from the countryside into London by train using new refrigeration methods. Timing is all; in 1864 he had established the Express Country Milk Supply Company in London, and a year later rinderpest wiped out London's cattle. He invented the milk churn and developed means of chilling. Within 20 years he was conveying to London 30,000 gallons of milk. This was topped with a Royal Warrant which has been renewed by every monarch including the present Queen, and a knighthood. Colonel Arthur Barham set about improving the property, including the still-extant fountain pool with lead figurines of spouting cherubs and dolphins modelled by Gilbert Bayes in 1913. He built a new stable yard in Elizabethan revival style in 1914. As a territorial soldier he was recalled for military service, but was invalided out in 1917. Tragically Wilfrid, his eldest son, was not so fortunate.

The curving backbone of clipped undulating yew was planted by Arthur in the 1920s, its role changing as it progresses – division, backdrop, enclosure. Wisely, he had underground walls built that prevent the yew roots from sapping all the goodness from the other plantings. One area has a sun-dial set within a large circle with rays of bricks and flints (an imaginative alternative to planting thyme). The Four Seasons culminates the gardens as the visitor circles towards the Eagle Slayer statue.

In 1955, after nearly 10 years of lying empty, the newly married David and Catherine Barham commissioned local architect Dennis Brown. In 1959 they rationalised and reconstructed the house in its earlier Queen Anne elegance. In 1960, with all mod cons, they moved in to enjoy 43 years of making the Hole Park Estate a thriving farming and property business. The symmetrical setting of the façade – gravelled promenade, steps to grassy terraces and a stone-edged pool – is softened by a magnificent lemon-scented *Magnolia grandiflora* and the yellow noisette flowers of trained Banksia roses.

In 2003 the current Barhams (Edward, Clare and their three children) moved into this estate which in 2020 employs more people than it did 50 years ago, thanks to its wide-ranging business activities in a glorious High Weald setting.

Borders, Hole Park

The Long Borders from the statue of the Eagle Slayer drift with colours and textures framed by close-clipped yew hedges. Beyond, a backdrop of trees and a cordyline whose spray of small white flowers looks like a fountain alongside the fastigiate yew. (Courtesy Hole Park)

(Following pages) Policy, rhododendrons, Hole Park

The woodland runs away on the left and the ground on the right rises to the house and lawns. A pleasing and colourful picture of microclimates with introduced conifers and rhododendrons contrasting with one another, as the eye is drawn through the area known as the Policy. Earlier in the year it has a fine display of *Camellia*.

FALCONHURST

amazing garden of flowers, and fields that go on for miles and look as if they connect with the sky

The highly successful Shrewsbury barrister John Chetwynd Talbot lived in style. As well as being a Victorian philanthropist, he built the church at Tixall, Ingestre near Shrewsbury. However, he decided to move his family from Shropshire to a house called Edells in Kent before buying the land on which Falconhurst now stands. He bought the surrounding farmland to create his own village complete with school, village cottages to house his staff and the teachers, and Holy Trinity church, modelled on the one at Tixall, whose parish which he named Markbeech, was taken from those of Hever and Cowden. An artist's impression survives showing a solid stone house set on generous terraces embosomed by trees in typical early Victorian style. The original front door makes an impressive garden arch, and some of the 19th-century delicate iron fencework and gates can be seen in the depths of the yew hedge. Tragically, he died from tuberculosis in 1852 just as the house was completed, leaving his wife Caroline to complete the project. Their son George enlarged the house considerably but in the early 1960s Tom and Cynthia Talbot decided to take this down. Today's gardens make full use of the ruins, the surviving architecture and the gentle trace of the original sweeping drive. You know you are on Talbot ground if you spot buildings sporting pigeon-blue paint.

Ruins to High Weald, Falconhurst
The pleasure of ruins where nature can be allowed a certain amount of free rein, filled with naturalised grasses formalised by a green path to a former window that frames the terraces. Rather than a roof the walls are topped with a view over the High Weald. (Courtesy Falconhurst)

Terrace, Lutyens seat, Falconhurst
Falconhurst's architecture is witness to the history of a Victorian status home, 20th-century adjustments, and the legacy of dramatic terraces. The pigeon-blue of the gutters curiously echoing the skies while the hydrangea flowers behind the Lutyens bench look like white party balloons. A setting of horticultural variety from lichen and moss to the textural delights of shrubs and trees. (Courtesy Falconhurst)

No family is an island, and friendships are built over shared estate problems, cricket or exchanging plants. The Meade-Waldos who sold Hever Castle to the Astors, were former owners of Stonewall Park just a few miles from Falconhurst. Evidence of their friendship with the Talbots is proven in a black and white photo taken in 1952 of a young Charles Talbot as page boy at a Meade-Waldo wedding.

In 1992 Charles and Nicola Talbot took over the 700-acre estate of farmland and woods. It is the perfect venue for film locations and weddings which they limit to eight a year, thus retaining its private family feel. This is in no small measure thanks to Nicky's garden philosophy of not aspiring to tidiness. Her grandchildren have their own splendid shepherd's hut tucked between the gardens and the safely grazing Jersey cows. The garden areas unfold with the views opening and closing in turn. There is a series of walled gardens with greenhouse, vegetables, fruit and large, flowering borders. One enclosure was converted for the Dig for Victory campaign during World War II, after which it was used for hens. Nicola

Walled garden, Falconhurst
Apart from shelter the walled garden has a treasure trove of old bricks here formed into a circle filled with flame tulips. The *Miscanthus* centrepiece grows into a screen that hides the seat as the summer elapses. (Courtesy Falconhurst)

(Following pages) Amphora, Falconhurst
An English garden in an Italian frame, irises and roses in close-clipped box beds. The recumbent amphora leading the eye up the steps to the pool framed by fastigiate yews. Beyond the sheltering walls the countryside takes over. (Courtesy Falconhurst)

cleared the area and created a formal stone-edged pool whose surface is decorated with waterlilies as well as reflecting the colourful planting around. Over the wall are the upper branches of a Japanese maple planted *c.*1910 which affords fiery colours in the autumn. The orchard has some stalwart heritage fruit – pears that ripen in November and wall-trained figs alongside a venerable wisteria; the old Peach House was a casualty of the 1987 gales. The gardener Marc Owen started in 2017 and is a follower of New Perennial gardening – grasses can be seen gently swaying around the swimming pool.

The earlier farm buildings on the approach to the house now host the Farm Shop; the former farmyard leading to an old pond had become a large rubbish dump, this is now sown as a wildflower area. Beyond the pond the view is over grazing for the mixed organic herd including pedigree Jersey cows and the traces of the old drive. Towards the house there is an excellent specimen of a strawberry tree (*Arbutus unedo*). Take time to sit in the Lutyens seat on the planted terrace for one of the best views of the High Weald towards Crowborough Ridge where wildness pleases.

CHAPTER TWO

The English park and woodland gardens

The English park, beautiful, sinuous and productive, through which one travelled awaiting glimpses of the mansion, was perfected and formulated by England's greatest 18th-century landscaper, Lancelot 'Capability' Brown. He is noticeably absent across the High Weald owing to a paucity of his usual patrons – wealthy nobility with extensive estates to improve economically and visually (not to be found in forests worked for modest returns). Elegance and prosperity were available in neighbouring Tunbridge Wells or on the coast at Hastings. However, as the 18th century closed a taste for sublime and untamed nature attracted visitors to the High Weald. William Cobbett's *Rural Rides* delightedly reported '. . . *as to human happiness, I am of opinion that as much, and even more, falls to the lot of the leather-legged chaps that live in and rove about amongst those clays and woods ... I have always observed that woodland and forest labourers are best off in the main. The coppices give them pleasant and profitable work in winter . . . And then in the great article of fuel!*'[1] Later in the 19th century one rail traveller wrote '. . . *within thirty-five miles of London you get a country as primitive and almost as lonely as that of the Adirondack . . .*'[2] By 1783 Humphry Repton had set up in business as Britain's first landscape gardener taking a picturesque approach, and he had several contacts in the region. It was the arrival of wealthy Victorian entrepreneurs, a taste for the picturesque near London and the passion for collecting that saw the making of the High Weald.

English parks and woodlands such as Sheffield Park, Wakehurst Place, Bedgebury Pinetum, Scotney Castle, Nymans, Stonewall and Ashdown Park, as well as the plantsman's paradise of Borde Hill are thriving witnesses. These magnificent landscapes survived two World Wars, only to be fragmented 40 years later by the Great Storm of October 1987 when some 15 million trees fell across Southern England. More than three decades later, many reflect that the obliteration created a new canvas for English parks and woodlands.

Lake, loggia and church, Ashdown Park
The lake is fed by a diverted stream from Wych Cross and animated by the fountain, across the lawns the large comfortable garden loggia and the church of Our Lady and St Richard built by the Order of the Sisters of Notre Dame of Namur. (Courtesy Ashdown Park House)

1. William Cobbett, *Rural Rides*, 2 vols (1853; reprint, Everyman's Library, 1966), p. 58.
2. Quoted in Peter Brandon, *The Kent & Sussex Weald* (Phillimore & Co., 2003), p. 218.

WAKEHURST

a cornucopia of plant hunting on a garden scale

In one day at Wakehurst the visitor can travel into the High Weald's prehistory along the geology of the Rock Walk, trek the Himalayan Glade, and stroll beneath the canopy of North American arboreal giants. If the spirit of adventure holds no allure, then the landscape and lakes nearer the mansion offer a cornucopia of plant hunting on a garden scale. The knowledge and experience of Kew is manifest throughout in the exuberant enthusiasm of displaying a world of plants; to learn where it all starts voyage in and around the building that houses the Millennium Seed Bank. The human history of the site is equally all-encompassing: from the Roman road that ran nearby, to the first owner who arrived with William the Conqueror, and Sir Edward Culpeper who built the original mansion in 1590 using Ardingly Sandstone and Horsham tiles (in fact, stone slabs), starting a century of Culpeper occupation. After two changes of ownership a planter of exotic trees arrived in 1869, the Dowager Marchioness of Downshire, who chose to plant the recently introduced *Sequoiadendron giganteum*. A grove of redwoods forms the entrance to Horsebridge Wood, joined by a *S. giganteum* which was raised from seed collected from the General Grant tree in California, now an apt part of the North American woodland area.

A subsequent change of hands led the consummate planter Gerald Loder (later, 1st Lord Wakehurst) to transform the estate between 1903 and 1936. One of the extraordinary family of gifted plantsmen as outlined in Chapter One (pages 19–43), he helped to sponsor plant-collecting expeditions in East Asia, the temperate southern hemisphere, and for conifers in general – exploring Wakehurst's Himalayan Glade is to walk in his footsteps. He chose the Westwood Valley area for its moister, cooler microclimate, knowing it to be ideal for rhododendrons. It leads into the Himalayan Glade with its wide range of Asian plants. On his death, Sir Henry Price safeguarded and continued his legacy, eventually bequeathing Wakehurst Place (as well as a sizeable endowment) to the National Trust, to whom it still belongs, in 1963. Loder had helped the Royal Botanic Gardens, Kew during their earlier tenure of Bedgebury Pinetum from 1925. In 1965 Kew leased Wakehurst, and there is still strong collaboration between Bedgebury, Kew and Wakehurst. The 1987 storm radically changed the landscape with a loss of 15,000–20,000 trees but provided the opportunity and impetus to tailor the redevelopment more closely with Kew's interests in science, education and conservation. This is manifest in the inclusion of native trees alongside the greater interest in woodland gardens, the younger introduced plantings have been and continue to be sourced from overseas collecting trips. They are grouped together according to geographical origin for both visual pleasure and scientific study.

Starting at Coates Wood, a full circuit back to the mansion covers approximately 3.6 kilometres (2¼ miles). The terrain is sometimes very steep, with height variations of 70 metres. Coates Wood takes you from the trees of Australasia, including *Wollemia nobilis* (discovered in 1994 by David Noble deep in the Wollemi National Park in the Blue Mountains 200 kilometres from Sydney), to those of South America. The soft autumnal oranges of Chile's contribution, *Nothofagus nervosa*, are an interesting contrast. Bloomers Valley stretches ahead with the Rock Walk on your left sometimes obscured by trees until it reaches the Sandstone Outcrop. Making up over three quarters of the world's botanical diversity, cryptogams (literally, hidden reproduction) do not flower or set seed, but spread by spores – ferns, mosses, lichens and liverworts. Their luxuriant growth on the sandy rock habitat at Wakehurst was first identified by Francis Rose – hence the Francis Rose Reserve leading to the Rock Walk. They are also indicator plants for clean air. Above the Rock Walk is Bethlehem Wood with noted collections of birch and beech.

Early summer pool and bridge, Wakehurst Place
Wakehurst is where the wildness pleases underscored by the astonishing global range of nature's arboreal and floral beauty. Away from the mansion visitors can roam under the canopy of tree giants, climb steep inclines, and enjoy the reflections of streams and pools.
(Wakehurst, Jim Holden © RBG Kew)

Symplocus paniculata, Wakehurst Place
The small, white, fragrant flowers of *Symplocos paniculata* appear in May and June. If there are two or more shrubs, plus a hot summer, brilliant ultramarine blue fruits follow in the autumn which usually persist into winter after leaf fall. A native of Taiwan as well as the Himalaya region, China and Japan.

A native canopy of oak and beech, with bluebells underfoot in the spring, heralds the entry into Horsebridge Wood and a North American landscape of redwoods – the giant *Sequoiadendron* and the dawn redwood *Metasequoia glyptostroboides* with its characteristic soft trunk. The area is dotted with Douglas fir, *Pseudotsuga menzesii*, known as a fragrant home for wildlife. In Scotland red squirrels and pine martens, in England birds and bats will shelter in deadwood cavities, whilst buzzards, sparrowhawks and hobbies use mature specimens as nesting sites. The seeds are eaten by smaller birds such as finches and small mammals, and the spruce carpet and dwarf pug moths feed on the needles. David Douglas introduced the tree in 1827. There are also *Liquidambar*, the American sweet gum, whose autumn colours are spectacular.

The walk towards Westwood Lake flanks a Site of Special Scientific Interest where no intervention is allowed. Two climbs give you spectacular views over the chain of lakes that follow from Westwood before passing through the Himalayan Glade to either the Westwood Valley or the Pinetum. The latter lost 80 per cent of its trees in the 1987 storm, and is being developed as a mixed North American and European landscape. The Slips leads you back towards the mansion through an eclectic array of trees chosen by Gerald Loder. Here green-winged, Jersey and other native orchids have naturalised. The Balustrade at the top has a sun-dial erected in the memory of Loder and his head gardener, Alfred Coates. Inscribed is a quote from the American Quaker poet John Greenleaf Whittier:

Give fools their gold and knaves their power,
Let fortune's bubbles rise and fall,
Who sows a field or trains a flower,
Or plants a tree, is more than all.

The Slips leads you to the Water Gardens and Iris Dell, not just Japanese irises, *I. ensata*, but giant Himalayan lilies, blue poppies and candelabra primroses, too. A traditional walled garden lies behind the mansion and a winter garden is maturing en route.

Compared with China and Japan, Korea has attracted few plant collectors; there are a small number of endemics but many hardy forms of other Asian species. The Tony Schilling Asian Heath Garden, specialising in plants from both Korea and Taiwan, curves away from the mansion. In the 1970s and 1980s botanical exploration by Kew scientists introduced Korea's lesser known plants such as *Sorbus commixta, Betula davurica, Quercus mongolica, Magnolia sieboldii, Rhododendron dauricum* and *Malus baccata*. 'Chinese' Wilson was one of the few who visited Formosa (now Taiwan); he collected *Rhododendron pseudochrysanthum*, which can be seen here. Other Taiwanese introductions include *Gaultheria itoana, Rhododendron nakaharae, R. pachysanthum, Skimmia japonica* subsp. *reevesiana* and *Sorbus randaiensis*. A visually pleasing end note for botanist and general visitor alike which sums up Wakehurst.

Sheffield Park **(see pages 63–65)**
Sheffield Park has all the elements of a traditional English park as a canvas for the owners' tastes and travels. A series of lakes surrounded by mature oaks and beeches plays host to a kaleidoscope of *Rhododendron* and *Acer*. Arthur Soames was the consummate plantsman, constantly seeking plants for Sheffield. Here, on Upper Woman's Pond, are the pads of Latour-Marliac water lily hybrids punctuating the autumnal reflections. Soames opened the grounds every day from early May to mid-November in aid of the County Hospital and the Nursing Fund. (GSW Photography/Shutterstock.com)

(Following pages) Early summer walled garden borders, Wakehurst Place
The sheltered walled garden lies to one side of Wakehurst Place mansion. Summer borders at their zenith shimmering in the sunlight. (Wakehurst, Jim Holden © RBG Kew)

SEEDS

To achieve an exact replica or clone of a plant it must be propagated by cuttings or layering. Raising a plant from seed preserves its genetic heritage. After the initial seed leaf, or pair of leaves, appears the plant produces what is known as the first rough leaf stage, the first indicator of their parental genes. At this stage hybridisers can hazard a guess at which genes are dominant. Further crossings are undertaken to eliminate unwanted characteristics or encourage a potentially more productive/attractive plant. In contrast to this and of importance in plant breeding are landrace varieties of plants which have developed in situ, evolving to adapt to their surroundings. Heirloom seeds will have been grown by the same family for generations, while heritage seeds have a traceable history. Both heirloom and heritage seeds are of great importance for the future viability of gene diversity.

Early summer meadow, Wakehurst
An early summer view across the naturalised ox-eye daisies that form one of the many meadows at Wakehurst. The low stone bridge and pool beyond invite the visitor to explore further.
(Wakehurst, Jim Holden © Kew)

Seed raised, *Xanthocyparis vietnamensis*, Bedgebury Pinetum
Seedlings of the Vietnamese golden cypress (*Xanthocyparis vietnamensis*), which can only grow on steep limestone mountains within a narrow climatic and altitude zone of northern Vietnam and southern China. It has clusters of small knobbly cones approx. 1cm when mature. As the young leaves mature, they flatten, fatten and become more frond-like. They are threatened by deforestation, logging and acid rain in the wild and then what will happen if the temperature rises? These seedlings are being raised at Bedgebury Pinetum with a view to reintroduction.

The Millennium Seed Bank at Wakehurst is the most biodiverse place on Earth – two billion seeds banked from over 40,000 wild plant species from over 100 countries in the MSB partnership. Architecture has a role to play; the MSB building runs solely on renewable energy from the photovoltaic installation on the roof, providing enough electricity to refrigerate the entire seed collection contained in its vaults. The room is kept at 15 per cent relative humidity and 15°C. These dry conditions are necessary as up to half the weight of fresh seed at harvest is water. If this was frozen it would kill the cells, so in order to store them successfully and increase longevity by around fortyfold, they are dried until they reach a moisture content of six to seven per cent. Dry cool air enters via an air inlet and flows around the seed collections. The warmer, moister air extracted from the seeds is removed via an air outlet to be re-dried. Scientists enter and leave via an airlock system thus limiting the moist air that might reach the room. Establishing the quality of the seed is very important, so small samples are X-rayed to check that the embryo is healthy and there is no insect damage.

BEDGEBURY PINETUM

not an arboretum, a botanical garden

For those who like their landscapes action-packed, Bedgebury National Pinetum has the National Cycle Network Route 18 running along its northern borders, with bike hire and parking for horse boxes. Within Bedgebury Forest you can choose from a 13-kilometre single-track MBT (Mountain Bike Trail), Go Ape, or two family cycle trails of nine and 4.5 kilometres, respectively – fitness and physical wellbeing guaranteed. What about the trees, bird song and scent? Hard to appreciate at speed, so it is worth slowing to walking pace: for a well-trodden route there are the three-kilometre Pinetum and two-kilometre Play Trails, but this large landscape copes with walkers going off-piste to look and listen. A Gruffalo-themed adventure awaits families, or for those prepared to roll up their sleeves, you can volunteer for a Wednesday Wild Crew and Sunday working groups. Twelve botanical artists have formed the Bedgebury Pinetum Florilegium Society,

***Pinus sylvestris* trunk, canopy and sky, Bedgebury Pinetum**
The expressive bark on the trunk of a *Pinus sylvestris* curving into its canopy. Despite being the 'Scots pine' it is very much viewed as the original English pine in the landscape.

Seed raised, *Callitris rhomboidea*, Bedgebury Pinetum
The Oyster Bay pine (*Callitris rhomboidea*) is a highly ornamental Queensland native that has naturalised in other parts of Australia

Seed raised, *Cryptomeria, Magnolia*, Juniperus, Bedgebury Pinetum
Bedgebury holds the National Collection of both *Cryptomeria* and *Juniperus*

invaluable as photography is, watercolours seem more effective in capturing the animation and characteristics of plants. Although pinetum is an umbrella term for a plantation of conifers, at Bedgebury one-third of the trees are not conifers; there are also lakes, acid grassland and heathland, in all they provide a canopy for wildlife such as bats, dragonflies, butterflies and an array of birds and ducks.

The stump of the Old Man of Kent mirrors the recent history of the site: the last owners of the Bedgebury Estate, the Beresford family, planted the Northwest American grand fir (*Abies grandis*) in 1880. The Forestry Commission was founded in 1919, and six years later, in 1925, bought the estate which it ran jointly with Kew until 1965. The 1987 storm wrought its vengeance on the 51-metre-high Old Man of Kent, taking one of his four primary limbs, and Storm Katie in 2016 took another. The final chop came in August 2016, leaving just a large stump. Fortunately a Bedgebury team had returned to Northwestern America to collect seeds in 2015; these have germinated, and the saplings are being planted in the Pinetum in 2020.

The original grouping by taxonomical order is gradually being changed, and new plants are being raised from collected seed rather than cuttings. On the northern boundary a quarter-acre (approximately 1,000 square metres) site is divided into 100 experimental plots which, in the past, have been used to study eucalpyts and the commercial possibilities of monkey puzzle trees (*Araucaria araucana*). Bedgebury is a member of the Global Trees Campaign run by Fauna & Flora International, and the International Conifer Conservation Programme (ICCP run by the Royal Botanic Garden Edinburgh). In the greenhouses (not open to the public) some of the world's rarest conifers are being raised from seed, not least the Mulanje cedar (*Widdringtonia whytei*) – now classed as critically endangered in its native Malawi – and

Seed raised, *Larix griffithii*, Sikim larch, Bedgebury Pinetum
The Sikkim larch (*Larix griffithii*) was first described by Joseph Hooker, later Director of Kew, in 1854. A deciduous conifer, it ranges from the Allegheny Mountains to northern India.

Seed raised, *Rhododendron catawbiense*, Bedgebury Pinetum
There are 800 species of rhododendron, of which four are known to be invasive. Here a seedling *Rhododendron catawbiense* in the greenhouses at Bedgebury Pinetum from the Allegheny Mountains in southeastern USA. It is a parent of many hybrids and related to the UK rhododendron nightmare *R. ponticum*. Each plant of the latter produces a million seeds a year, the branches root and it now covers 3.3 per cent of Britain's woodland. Annually Wakehurst remove 50 square metres.

the Chilean plum-fruited yew (*Prumnopitys andina*). Bedgebury was the first to successfully germinate the seriously endangered *Xanthocyparis vietnamensis*, Vietnamese golden cypress. They have had equal success with the Chichibu birch (*Betula chichibuensis*). As a global resource, conifers provide building materials, cardboard, furniture, medicine, musical instruments and paper, ideally from well-managed sustainable forests with a PEFC or FSC logo.

Bedgebury holds the national collections of *Chamaecyparis lawsoniana*, *Cryptomeria japonica*, *Juniperus*, *Taxus*, and heritage cultivars of *Thuja* and the thuggish x *Cuprocyparis leylandii* cultivars. The Visitor Centre has a series of imaginative information panels, an excellent café and delightful outlook across the lake towards the Pinetum. The large presence of conifers means their shapes and colours in the evening light are highly visible and makes this an attractive landscape.

Play up, and play the game, was a sentiment urged in all parts of the British Empire. To this day immaculately kept grounds can be seen in regions with the most unpropitious climate and/or soil. Across the High Weald many local cricket grounds were started within large estates in the years leading up to World War II, the team comprising employees from 'the big house' playing against the village team. Sheffield Park has the most illustrious history: Henry Holroyd, 3rd Earl of Sheffield (1832–1909), had a lifelong enthusiasm for cricket. As a 13-year-old he took part in the first recorded cricket match at Sheffield Park in August 1845. He went on to play for the Gentlemen of Sussex. In 1876 he inherited the estate and set about creating a cricket pitch to the south-east of the Upper Woman's Way Pond. This also provided much needed paid winter employment at a time of agricultural depression. Six years later he put up an octagonal pavilion set in its own garden, around which he planted climbing shrubs, a separate ladies' pavilion was added later. The origins of cricket matches between England and Australia can be traced to Sheffield Park – hosted here between 1884 and 1896. The Australian team would open their tours with a match against Lord Sheffield's XI. The final match was attended by the Prince of Wales, later Edward VII, and a remarkable 25,000 spectators crowded in, free of charge, to watch – the philosophy being that if it was free they would not pinch the plants. Lord Sheffield's team was captained by one of the greatest cricketing heroes, W.G. Grace and included Prince Ranjitsinhji, C.B. Fry and F.S. Jackson – the match was drawn. During 1891/2 Lord Sheffield went to Australia to promote the English team led by W.G. Grace, this culminated in the domestic first-class cricket competition of Australia for the Sheffield Shield, donated by Lord Sheffield.

(Previous pages) Bowling green, Gravetye
The English garden is celebrated for its greens because the climate suits bowling green grass. The English weather calls for shelters, here at Gravetye Manor in arts and crafts style. The garden pavilions are built in local stone with Horsham tile roofs – the immaculate grass is good for bowls or croquet. The backdrop could be described as hardy and global. (Courtesy Gravetye Manor)

Back on the High Weald, Nymans' car park was where the locals of Handcross played while the Nymans XI played further down the hill. High Beeches played below the present house where the spot is still known as 'the cricket field'. In Justice Bowen's time, Colwood had an estate ground behind the mansion while the village of Warninglid used to play behind the village stores.

A quote by John Chetwynd Talbot (who founded the cricket club a couple of years later) relates to a celebration of the centenary of Falcolnhurst: '*The Match at Falconhurst on August Bank Holiday reminded me that in more spacious times – and particularly before World War One – country house games were a big feature of cricket in Kent.*' Between 1888 and 1894 Falconhurst hosted women's cricket as well as celebrity umpires such as Stanley Baldwin, later prime minister and cousin of Rudyard Kipling at Bateman's, and Freeman Thomas, later Viceroy of India. Following a visit to Falconhurst, Walter Hardinge, the Kent county cricketer 1902–1933, got his chance to play. At the match in the opening quote, two bats were broken but the family side won with 10 minutes to spare and five wickets in hand. Not just a cricket tea followed but a centenary cake with a coloured marzipan replica of the house was served.

Cricket grounds dominate the surroundings of these High Weald gardens where wildness has been controlled, but the immaculate grass of bowling greens that often double as croquet lawns should not be forgotten. Nor yet the more active lawn tennis, all requiring patient and diligent attention. None of the above exists at Bateman's, just a record of the quarry garden being used as a theatre. However, steely competitiveness disguised as leisure in these games was captured by Kipling in 'The Islanders' (1902):

Soberly and by custom taken and trained for the same,
Each man born in the Island entered at youth to the game –
As it were almost cricket, not to be mastered in haste,
But after trial and labour, by temperance, living chaste.
As it were almost cricket – as it were even your play,
Weighed and pondered and worshipped, and practised day and day.

. . . .

Will ye pitch some white pavilion, and lustily even the odds,
With nets and hoops and mallets, with rackets and bats and rods?

Cricket match, Falconhurst

At Falconhurst a centenary cricket match was played between the club and the family team in 1952. The Talbot family team won with the game making two full columns of the local paper. Cricket remains an integral part of the English country scene, the big house team against the village with a slap-up tea afterwards. (Courtesy Falconhurst)

NYMANS

And with rich gifts of tree and shrub and bloom
Gladden the gardener's heart, the artist's eye.

From 'A Benison' by Sir George Cockerill, dedicated to Maud Messel

Nymans is organised to please. The Visitor Centre does not impinge on the garden but ensures that visitors are informed and refreshed before stepping into its fascinating historic landscape. The National Trust invariably have a blackboard advising on what to do that day, often with delightful vases of flowers and leaves that are at their prime. Nymans is no exception, Bateman's and Scotney Castle do the same. Echoing the habit of Anne, Countess of Rosse (née Messel), who continued to stay here until her death in 1992, flowers from the garden are arranged in the rooms by a volunteer team. More than 30 plants here carry names associated with Nymans: for example, *Magnolia* 'Anne Rosse' and *M.* x *loebneri* 'Leonard Messel', *Camellia* 'Maud Messel' and *C. japonica* 'Leonard Messel', *Sorbus* 'Leonard Messel', and more recently *Eucryphia* x *nymansensis* 'Nymans Silver' and a fastigiate handkerchief tree, *Davidia involucrata* 'Nymans Sentinel'. One of the finest UK collections of Chilean plants grows here. Many specimens dating back to seed collected in the 1920s, they are mostly to be found in the shelter of the Wall Garden. An initial walk towards the pinetum and arboretum offers the opportunity to appreciate the expansive landscape that surrounds Nymans. Stop in the temple which celebrates the life and work of the German architect Alfred Messel (1853–1909). In 1890 his brother Ludwig Messel, a German banker, bought Nymans with 600 acres for his wife Annie and their six children, Leonard, Ottilie, Ruth, Harold, Hilda and Muriel. Ludwig commissioned his brother to enlarge the Regency house, to include a large conservatory and a tower overlooking the main lawn. In 1895 James Comber was appointed Head Gardener, a job he held until his death in 1953. Muriel was 10 years younger than Hilda, and she followed her father and James around the garden. In her teens she approached William Robinson at Gravetye for advice on creating the borders in the Wall Garden. In his last years Ludwig decided to emulate his friend and neighbour Sir Edmund Loder and make a plant list of all that was growing at Nymans. Poor health meant the project was abandoned but Muriel finished the catalogue, Robinson wrote an introduction, and the artist Alfred Parsons who painted the pergola at Nymans, illustrated *A Garden Flora* which featured more than 2,000 plants at Nymans. It was published in 1918.

In 1915 Leonard inherited, and his wife Maud wanted the house redesigned. Finally, in the 1920s, they remodelled it in the style of a medieval West Country manor where their three children Linley, Anne and Oliver grew up. Leonard also inherited James Comber as his head gardener with whom he developed a strong friendship. They made a formidable team of plantsmen, regularly exhibiting at the RHS Shows notably with *Camellia*, *Magnolia* and *Rhododendron*. Their plantings of *R. macabeanum* introduced by Kingdon-Ward *c.*1928, and *R. johnstoneanum* introduced in 1934 can still be seen. James' son Harold (see page 62) sent seeds from his Andes expedition. These were cultivated and still grow in the Wall Garden: the Chilean spiny, fragrant *Discaria discolor* and the beautiful evergreen trailing *Asteranthera ovata*, which flowers in June. Maud looked for beauty, scent and romance in the garden and what might be picked to arrange in the house. Her friendships included Ellen Willmott who famously employed 100 gardeners and the indefatigable Norah Lindsay, both of whom visited Nymans.

Meanwhile Anne and Oliver became leading lights in London society. Anne was to be the last family occupant and Oliver the highest paid and most successful theatre designer in Europe, finally retiring to Barbados in 1966. On at least one Christmas Oliver redecorated the Great Hall in suitably dramatic style. During World War II it was used to house evacuees. In February 1947, with all the water pipes frozen, a disastrous fire swept through most of the building. Anne and her second husband Michael, 6th Earl of Rosse, helped to save as much as possible. Leonard and Maud realised that they were unable to rebuild so they bought nearby Holmsted Manor. During the 1950s Anne (1902–1992) and Michael used the house to entertain friends, including many from the

Ruined house façade, Nymans

Following a devastating fire in 1947, the façade of Nymans became a shell but one that encapsulates the pleasure of ruins. The array of trees and climbers with a small monkey puzzle tree in the foreground convey a wonderful sense of abandon. Totally false as everything is carefully orchestrated. (Stefan-Kadar/Shutterstock.com)

gardening world, earning the house the nickname 'The Potting Shed' albeit of rather grandiose proportions.

After Leonard's death in 1953, Nymans was given to the National Trust with 570 acres of wood and farmland plus financial endowment; Maud and Anne were given a life tenancy. A ceremony took place on 26 March 1954 with Vita Sackville-West announcing the opening of the gardens, and Ralph Stephenson Clarke of Borde Hill a guest of honour. James Comber had also died. Having worked with him for 30 years, Cecil Nice took over as the head gardener until 1980, turning the kitchen garden into a profitable concern. On Maud's death, Anne was appointed Garden Director. With her husband Michael, she oversaw further plantings of rare trees and shrubs. 'Plant big, darling' was the cry her son remembers as she searched for bold and beautiful plants for her flower arrangements. She resigned in 1987, after the Great Storm destroyed 500 trees including 40 champions. She died at Nymans in 1992, her former rooms are open to the public and remain as she left them. Harold Messel's son Alistair Buchanan took over initially, with a special interest in the Forecourt Garden during Anne's lifetime he now dedicates his attentions to the topiary on the Terrace and in the Davidia Glade. Having started at Nymans in 2015, Joe Whelan is now the head gardener, his experience and knowledge are wide-ranging, notably garden history and the care of Nymans' national collections.

The Pinetum was redesigned by Isabelle Van Groeningen and John Sales who framed the Wealden views with sequoias, cedars and redwoods as well as acers. I suggest following the lime avenue diverting into the Davidia Glade to admire Alistair Buchanan's topiary. On to the house façade and pause to admire the countryside to the east from the Prospect. The ruins of the house are a majestic backdrop of many parts. Heading to the southerly façade the clipped box hedges provide year-round formality. From the former tennis lawn are views to the classical loggia and the sunken garden that was such a feature of interwar gardens. The Main Lawn

Naturalised crocus, Nymans
Naturalised blue and white crocus form pools of floral colour under the trees at Nymans. They will be followed by narcissi, a few early flowers can be seen to the left. (Laurence Gough Shutterstock.com)

and 21st-century Winter Walk have large collections of dogwoods, grasses, hellebores and pulmonarias which provide colour and interest from winter into spring, with the scent of daphnes and witch hazel, notably *Hamamelis* x *intermedia* 'Rubin', wafting abroad. The Rock Garden created by Pulham and Son used the rocky spoil from when the croquet lawn was levelled. The Heath Garden is the largest of its kind with some 50 varieties of winter-flowering *Erica* and late summer-flowering *Calluna* interspersed amongst dwarf varieties of rhododendrons and pines, and small roses.

The magnificent wisteria-clad pergola leads back to the Main Lawn with its stately cedar of Lebanon. The trees reveal and conceal contrasting views of the ruined original house and the effectively untouched 'medieval' West Country extension. The house and its surrounds offer a variety of horticultural experiences, the walls provide a series of microclimates and sense of theatre. The most recent development has been the creation of the 'Garden in the Ruins'. Literally in the ruins of the Great Hall, it is inspired by Oliver Messel's dramatic style and designed by Nick Delves, Assistant Head Gardener. Trellis, water features and theatre flats form the scenery in which the leading, support and chorus roles are fulfilled by plants. Two thoughtful initiatives have developed from conversations with three local Alzheimer Society groups: firstly, for one hour after opening the area will be set aside for quiet contemplation, and secondly, planting has been selected that will enliven the ground whilst their scent will rise up, thus stimulating two senses. (Dementia sufferers seldom look up, preferring to cast their gaze downwards.)

The Wall Garden is the centrepiece of the Nymans horticultural excellence. Originally it was an orchard, but Ludwig Messel made it his trial ground with meandering paths to invite the curious. Later, Muriel Messel created the Summer Borders which continue to flourish alongside magnificent examples of the handkerchief tree, *Davidia involucrata*, and the *Magnolia campbellii* with flowers the size of dinner plates. The Chilean Border is key to the awarding of Plant Heritage status in 2016 to The Harold Comber Collection which consists of 60 taxa that are either original or propagated from originals. These include a form of *Desfontainia* and *Escallonia* x *stricta* 'Harold Comber' and the Plant Heritage gifted *Luma apiculata* 'Nana'. The collection has been supplemented by introducing newly collected plants from the same region, such as the Chilean endangered native *Myrceugenia leptospermoides* sent by Martin Gardiner of the Royal Botanic Garden Edinburgh.

The Rose Garden was created by Maud Messel who collected old and rare roses, often from cuttings taken in friends' gardens. The design was reworked in 1989 by Isabelle Van Groeningen, introducing more arches and pillars. The Rose Garden fountain, commissioned by Anne and Alistair Buchanan, was sculpted by Vivien ap Rhys Pryce. Take a walk on the wild side amongst Chinese and Asian trees collected by Forrest and Kingdon-Ward; the Wild Garden was first used as a trial ground for the plants raised from seed that had been returned from plant-hunting expeditions. The Rhododendron Walk includes the parent of many hybrids, the Himalayan *Rhododendron thomsonii*, and the Japanese/Chinese/Korean *Quercus acutissima.* The nearby Top Garden is the beating heart of both the gardens and the Plant Centre (which, in traditional style, also features the June Borders that bring in summer colour). Weather permitting, allow time to simply imbibe all Nymans has to offer.

SHEFFIELD PARK

The great ponds at Sheffield Place at the right season of the year are bordered with red, white and purple reflections, for rhododendrons are massed up on the banks and when the wind passes over the real flowers the water flowers shake and break into each other.

'Reflections at Sheffield Place' by Virginia Woolf

Capability Brown's earthmoving and hydraulic skills inspired and provided exemplars for many followers which probably accounts for the Brownian setting of Sheffield Park in East Sussex and its unsubstantiated attribution. Horticulturally the greatest legacy is that of successful Lincolnshire brewer Arthur Soames who acquired Sheffield in 1910, his plant collecting gives the park artistic dynamism. What we see today is the culmination of multiple layers of design and planting which weave together in a pattern of horticultural excellence across the landscape and are reflected in the lakes. For a landscape where sheep might safely graze, appropriately 'sheep clearing' is the derivation of the name Sheffield as recorded in the Domesday Book. Many well-known historical families owned the estate, such as William the Conqueror's half-brother Robert, Count of Mortain, and the Sackvilles. Sheep were replaced by deer until the land was turned over to coppice timber for charcoal to feed the thriving Wealden iron industry. John Baker Holroyd and his heiress wife Abigail Way bought out the debt-ridden 2nd Earl de la Warr in 1769 for £30,000. In 1781 he was created Baron Sheffield, and the 1st Earl of Sheffield in 1816. A politician, he was notably active in trade and finance. The estate benefitted from Lord Sheffield's success as an agricultural improver, part of which was the creation of the sinuous landscape through and around Sheffield Wood. Between 1789 and 1790 Humphry Repton made several visits, his picturesque principle was to create a foreground, that is, by the house, middle pastural ground against a background of mature trees and, ideally, hills. At Sheffield he created four small lakes, divided the lower lake with a bridge and planted

Sheffield Park lake

View across lake below the house to the bridge cascade in summer. The effect is textured and serene. The outlines of the trees offer a global contrast, notably the redwoods and oaks with undernotes of acer. (Golden Shark 2/Shutterstock.com)

Sheffield Park house and landscape
Gothic Revival Sheffield Park House, designed by James Wyatt, was sold separately in 1953. In his 1934 article Wyatt writes of *Quercus coccinea* 'Splendens', *Amelanchier, Nyssa* and Japanese acer, all of which he planted around the lake and across the landscape for autumn colour. Today there are young *Photinia* nearer the house, the original *Photinia variabilis* were raised from seed by Soames. (Guy.G/Shutterstock.com)

up the banks. Before falling out with Lord Sheffield he wrote : '*A very beautiful lake has been added to the scenery of a place which abounds in the most perfect specimens of the picturesque effects produced by Gothic architecture.*'

The South Park has a range of parkland walks enclosed by the River Ouse whose flood meadow is frequented by kingfishers. As a boy George Holroyd, the future 2nd Earl, sketched his father's 'Iron Gates' for the newly installed canal lock, deepening and canalising the river to enable navigation through the Weald including the Sheffield estate. The trade was mainly barges carrying lime, chalk, bricks or coal for local small industries or agricultural use. Its peak usage also expedited its downfall as the lock was used to offload some 19 million bricks for railway bridges and viaducts. Sheffield Park is the final station on the Bluebell Railway. The series of meanders evidencing the ancient course of the Ouse across the meadows is being managed so as to absorb the flow of heavy seasonal water in order to prevent flooding and protect Lewes, as well as encouraging wildlife.

Sheffield Park continued to mature under the 2nd Earl while his son, the future 3rd Earl, joined the diplomatic service and served as MP for East Sussex from 1857 to 1865. He drained the Ten-Foot Pond to enlarge it and created the Middle Lake by merging two of Repton's. In 1882 he commissioned Pulham and Son (famed for their artificial stone Pulhamite) to create decorative garden rockwork and stonework for Top Bridge. Through the 1880s the water landscaping, the planting of an arboretum (which included rare varieties brought back from his posting in Constantinople), and the help of his gardener William Thomas Moore all helped to establish today's view from the house over water terraces to the cricket pitch, where his greatest passion lay. He died in April 1909, without an heir, leaving the estate in debt.

In 1889 the wealthy Lincolnshire brewer Arthur Gilstrap Soames fell in love with Sheffield while staying nearby. He asked for first option to buy the estate on the 3rd Earl's death and was one of his principle creditors. In 1910 the purchase was completed. His plant passions were roses and rhododendrons. Action was immediate: starting by improving the soil using thousands of cartloads of loam (the colours so praised by Virginia Woolf were thanks to him), he planted rhododendrons in clumps, notably along the northern side of Ten-Foot Pond and Middle Lake, which act as mirrors to their glorious colours. He made the same cross as Lionel de Rothschild of *R. discolor* x *griffithianum* to breed his own *R.* 'Angelo' and planted his fellow enthusiast's *R.* 'Loderi'. Spring fireworks are matched by the autumn colour, mirrored in Ten-Foot Pond, of scarlet American oaks (*Quercus coccinea* 'Splendens'), the tupelo gum (*Nyssa sylvatica*) and the wonderful Japanese maple (*Acer palmatum*). Among the large conifers he planted were Wellingtonia (*Sequoiadendron giganteum*) and *Abies procera*, the noble fir. Soames wrote a detailed article 'Autumn Colours at Sheffield Park, Sussex' for RHS magazine *The Garden* in 1934. He spent his summers in the South of France, hence his focus on spring and autumn. In contrast to arboreal colour and height, Soames allowed the south-west section of the garden to be more open with wild heathland. However, long winding beds of autumn gentians lined the route from the Cascade Bridge. Like the Husseys at Scotney Castle (see pages 157–159), Soames was a *Kalmia* enthusiast, sourcing his plants from the Knap Hill Nurseries. He crossed plants to obtain better strains of deep-pink flowers, especially after receiving the 'best pink' from Professor Sargent of the Arnold Arboretum. Soames planted the Kalmia Walk around 1911.

In 1919 Soames married Agnes 'Nellie' Peel who in her youth had travelled to the Arctic through the Kara Sea and Siberia. Soames died in 1934; Agnes lived to nearly 100 and ensured the estate survived World War II. Arthur Soames' nephew Captain Granville Soames took over in 1949. Working with the then head gardener T.H. Setford, he restored the gardens to their pre-war standards. In Kent, his son Christopher Soames was helping his father-in-law Winston Churchill to farm and garden Chartwell. In 1953, Arthur Gilstrap Soames was obliged to sell, and the house and gardens were separated. With encouragement from the RHS, the National Trust acquired the gardens in 1954. To celebrate the Queen's Silver Jubilee in 1977, the Queen's Walk was created by the head gardener, Archie Skinner, followed by the Stream Garden. Thanks to his attentions, Sheffield Park holds the Plant Heritage Collection of Ghent azaleas, fragrant with honeysuckle-like flowers which bloom at the end of May. The Great Storm of 1987 ripped out more than 2,000 trees; the woodland plants that had thrived in their shade gradually disappeared. Major replanting of native and exotic species was undertaken, including the palms now flourishing around Ten-Foot Pond. However, the most exotic planting at Sheffield is its autumn colour, with *Nyssa* being the signature plant. The lesson learned is: don't be afraid to take out oversize trees. The signage here is perfect, wooden finger posts give you all the information without impinging on the scenery, the paths are gravelled firmly in a multiplicity of directions.

Rhododendron undercanopy, Sheffield Park
As rhododendron plantings pass their centenary the undercroft takes on a sculptural form. However, they are space hungry and can be thuggish. Many of the original specimens are being revived and renewed with careful management. (Sue Martin/Shutterstock.com)

STONEWALL PARK

And then my heart with pleasure fills,
And dances with the daffodils.
The Daffodils, William Wordsworth

Stonewall Wood provides a typically High Wealden canopy of oak and beech. In 1968 Elizabeth and Valentine Fleming found Stonewall as it was about to be demolished, but with their four sons they breathed new life into it. Along with the house, the purchase included two cottages and 140 acres. On the edge of the park is a large immaculately maintained cricket field for Stonewall Cricket Club, one of their sons, Matthew, went on to captain Kent County Cricket Club. The old park drive has been superseded by the more direct Lime Avenue with each tree pruned into a cone, planted in memory of Valentine in 2015. The ground falls away to dramatic rhododendron grounds on the left while the walled garden lines the drive on the approach to the house. One of the architectural delights is the *c.*1850 camellia conservatory attached to the house; the wooden roof has now rotted but the metalwork is intact. It is a popular location for photographic shoots.

In the 16th century the Woodgate family owned land in Chiddingstone and Penshurst including Stonewall. In the early 19th century John Woodgate made his fortune as a banker in Tonbridge, rebuilt the house, and laid out the park. (Interestingly, the parquet flooring in the house looks like the intricate inlay of Tunbridge ware.) He put in the herringbone-brick ha-ha to open out the park views; a gap in the trees afforded a glimpse of Penshurst. His tenure was short-lived. When the Tonbridge Bank failed, amongst many country banks in the 1820s, he was forced to sell. Edmund Meade-Waldo had inherited Hever Castle in 1841 but had no desire to live there, so he leased it out as a working farm. He bought Stonewall and settled with his wife Harriette, creating the Pleasure Grounds as well as planting thousands of daffodils which have naturalised. Stonewall was managed by their son, the great ornithologist Edmund Meade-Waldo, until his death in 1934, followed by his sister 'Aunt Meg' Hay-Drummond-Hay nee Meade-Waldo who allowed local children to jump in the lakes. (She was remembered as something of an eccentric when she died in 1954.) The estate was sold in lots from 1966.

The Flemings employed David Chilcott as the gardener. He still cycles seven miles from Edenbridge and back every day. One of his ongoing tasks over the decades has been to control the thuggish *Rhododendron ponticum* among the prized mature species rhododendrons, and to clear the extensive Valley Garden which was choked with brambles. Elizabeth and Valentine, helped by David, added new species and created a woodland garden as well as opening out views, the water courses through to a large lake acting as a mirror to the trees and rhododendrons around it. The daffodils in March and Bluebell Glade in May draw in large crowds annually.

Parts of the original house are built into the brick walled kitchen garden and coach house west of the main house. A formal garden and pool frame the entrance from the house. It is laid out as a flower garden with herbaceous borders and lawn inhabited by a fearsome scarecrow. The main path is lined with gnarled espaliered fruit trees; other fruits include figs and a mulberry tree . . . and yes, the family are related to Ian Fleming.

Façade portico, stone dogs, Stonewall Park House
Stone dogs guard the porticoed entrance to Stonewall. Beyond is a glimpse of the ornate camellia conservatory which has proved popular for photo shoots.

(Following pages) Sheep and cricket ground from ha-ha, Stonewall Park
Elements of the English landscape – a ha-ha allows an unimpeded view to the sheep in a pasture with an ancient oak. In the distance the grass becomes well-tended for the cricket pavilion of the Stonewall Cricket Club. The cricketer Matthew Fleming grew up here, he was Captain of Kent County Cricket Club and served as President of the MCC (Marylebone Cricket Club) in 2016–7, he returns to play on this his home pitch.

ASHDOWN PARK

This blessed plot, this earth, this realm, this England
This nurse, this teeming womb of royal kings
Richard II, William Shakespeare[1]

Sporting activities date back to when Ashdown Forest was a royal hunting ground. Within the Park's vicinity today are a golf course, putting green, tennis court, croquet lawn, no less than three jogging trails and a llama park. The first mansion house was built in the early 19th century when the estate comprised 3,563 acres. In 1867 the new owner Thomas Charles Thompson, MP for Durham, demolished and rebuilt the house. The grounds were redesigned: the stream from Wych Cross was dammed to create water features that lead to the

Light-footed Hebridean sheep from Ashdown Forest
Where sheep may safely undertake conservation grazing and help wild flowers flourish. Chosen for their light small hooves, Hebridean sheep from the Ashdown Forest are brought in during the autumn and winter to tread the ground, exposing bare soil where wildflower seed can transform a grass field. Bloomers Valley at Wakehurst Place is part of an initiative to restore traditional hay meadows which typically have 50 plants per square metre supporting insects, birds and other animals. This is achieved in two ways: by spreading species-rich hay supplied by the Beech Estate, Battle from the Ashburnham Meadow, and from local-origin seed supplied under the Weald Meadows Initiative.

1. John of Gaunt's speech. In 1372 Ashdown Forest was granted as free chase to John of Gaunt by his father Edward III

Croquet lawn, Ashdown Park
From the rough to the smooth, croquet lawn with steps to the sunny terraces by the mansion. The trees, such as the unmistakeable outline of the monkey puzzle (*Araucaria araucana*) are now mature specimens.

lake, ornamental trees were planted, some of which are now mature specimens. Although Thompson's daughter inherited on his death in 1892, she and her husband never lived in the house, nor did their son, who was killed in 1917. The house was used as a hospital and convalescent home for Belgian Army officers during the war. In 1918 the estate was purchased by the Oxcroft Colliery Company who promptly felled most of the trees for pit-props before selling it a year later to the Order of the Sisters of Notre Dame of Namur, Belgium as a training convent for novices.

The order was richly endowed by the Towneley Estates and the Honourable Mrs Edward Petre, enabling the house to be considerably enlarged with the addition of the West and East Wings as well as the beautiful church which was consecrated in 1927. For three years from 1971 to 1974 the United States International University took over, before it became a Management Training Centre for Barclays Bank. The current owners bought the mansion with 186 acres, of which 95 are landscaped woodland, and 40 grazing land with wild deer, leaving just over 50 for pleasure and leisure.

On arrival visitors are greeted by the 'Singing Lily' fountain, which was commissioned by Ian Gill in 2018, to celebrate 25 years of Ashdown Park Hotel. There is a lily for each year. Tucked into this mature landscape are the walled garden, sunken garden and secret garden, as well as the Bluebell Wood in the Eighteen Acre Wood.

CHAPTER THREE

Arts and crafts and the English country garden

'*Another thing, also much too commonly seen, is an aberration of the human mind, which otherwise I should have been ashamed to warn you of. It is technically called carpet gardening. Need I explain it further? I had rather not, for when I think of it, even when I am quote alone, I blush with shame at the thought.* William Robinson quoting William Morris in the frontispiece epigraph to *The English Flower Garden* in 1883. Morris had spoken these words in a series of lectures titled *Hopes and Fears for Art* in the 1870s. They were finally printed in 1882. Robinson had published his first book *The Wild Garden* in 1870, a key message being that planting should be used to reflect the garden's relationship with the landscape, contrasting less as it gradually meshed with the local flora. *The Wild Garden* was inspirational for the great English gardener Gertrude Jekyll who would become a friend and colleague. In 1871 Robinson had founded *The Garden* magazine (one of eight instigated by him), which he also edited until 1899 when Gertrude Jekyll took over.

Arts and crafts principles defined by Morris and his followers can be summarised as the use of local materials such as stone, wood and brick, and the use of hardy plants with purpose. They believed that the garden should be used as inspiration and illumination in tapestry, painting and literature. They extolled the medieval practices of working with the materials that surround you and using craftsmen rather than factory-made items. In all, a counterblast to the industrialisation celebrated by Prince Albert's Great Exhibition of 1851.

In 1856 Morris began work in the architectural office of G.E. Street where he met Philip Webb. Morris soon abandoned architecture to become an artist, craftsman and writer, and within four years he commissioned Webb to design Red House, Bexleyheath. Morris' first wallpaper 'Trellis' was inspired by the rose trellis in its gardens and can be seen in the Dog Leg Corridor at Standen. Known as the 'Father of Arts and Crafts Architecture' Webb provided practical common sense that matched Morris' romantic utopianism. He became a partner in Morris, Marshall, Faulkner & Co., co-founded SPAB (The Society for the Protection of Ancient Buildings), and wrote its Manifesto with Morris in 1877. They called themselves 'Anti-Scrape' on the basis that key to protection was stopping the crude restorations undertaken by other Victorian architects. Webb provided practical solutions that restored without destroying surface texture, thanks to him, the Ancient Monuments Protection Act was passed in 1882. Webb had strong High Weald connections: he designed Standen for the Beales in 1891 (see pages 80–82) and retired to Caxtons in the parish of Worth where the restorations of its beautiful church attracted his ire. He was equally damning about works on Burwash church, near Bateman's, describing them as 'witless fiddling'. Caxtons is a fine example of the High Wealden vernacular, a half-timbered house with red brick infill and weatherboarding on its upper storey. Webb welcomed visits from young architects, one of whom A.R. Powys (1882–1936) was also a pioneer in the conservation of ancient buildings. He wrote essays on architecture and *Repair of Ancient Buildings* in 1929, and was Secretary of SPAB before and after World War I. He helped with the conversion of the house at Sissinghurst and designed the curved wall in the Rose Garden there.

In the context of both the High Weald and arts and crafts in the English garden, the prolific and opinionated writer William Robinson is the key figure. Irish by birth he travelled regularly to France and to the United States, he promoted North American hardy trees, shrubs and flowers. Robinson's many writings and books, when sifted of some of the polemic on housekeeping and open fires, remain inspirational today.

Flowers around Gravetye Manor
The arts and crafts ideal of a house growing out of its own site, Gravetye Manor afloat in a sea of flowers, all the hard landscaping is in vernacular materials. In William Robinson's time even the grass was filled with flowers, edged in fragrant *Dianthus*. (Gravetye Manor)

GRAVETYE MANOR

where wildness and productivity please

From the 1520s Wealden ironmaking expanded rapidly, peaking between 1610 and 1635. Richard Infield worked Mill Place Furnace. He was married to Katharine Compton of Hampshire, and they had nine children. Their son Richard married Katherine Culpeper, daughter of Sir Edward of Wakehurst Place. Gravetye Manor was their great status symbol, a notable Elizabethan stone mansion built in 1598. In 1885 William Robinson bought the estate with 700 acres and started writing his *Tree and Garden Book*, a meticulous and opinionated journal of the work on the estate, house and gardens, later published as *Gravetye Manor or Twenty Years' Work round an Old Manor House*. As a pioneer of the natural garden the topography of Gravetye worked well; he cleared the sandstone rocky outcrops, harnessed the old mill streams, and converted the disused marl and mine pits, and hammer pond into lakes and pools.

His Gravetye notes start on 24 August 1885: under 'Wild Gardening' he introduces his revolutionary approach to daffodils – common, Poet's, Tenby and Star – as well as wild tulips (*Tulipa sylvestris*), which were planted not as 'tile-work' but in 'free natural looking colonies' under trees and in meadows and left to naturalise. In time he did the same with snowdrops, crocus and scilla, extending the idea to 'Spring flowers in turf' a suggestion that can be adapted to smaller gardens. He recommended varieties of the bulbs above plus *Hepatica*, *Anemone blanda*, winter aconite, snakeshead fritillaries and the rather larger *Camassia*. Although he does not record how many he planted in 1885, the greatest totals are for daffodils in 1896 (100,000) and 1897 (80,000) along the margins of the Lower Lake, and 1904 (50,000) plus regular plantings during the last 30 years of his life.

Before we explore Gravetye today a few quotes capture his drive. Observation: '*I am never concerned with Claude but seek the best expression I can secure of our English rural landscapes, which are far finer*' and '*Shelter is a great aid, and recesses in shrubberies or in banks clothed with foliage, form the most fitting background for beds or groups to nestle in.*' Opinion: '*In no case should regular steps be permitted in or near the rock-garden. Steps may be made quite irregular, and not only not offensive to the eye, but very beautiful, with violets and other small plants jutting from every crevice.*' And in the 'Virgins Bower' on clematis: '*A garden is of all things in the world, the place to select. In botany all plants are of equal value; but in gardens we must choose or suffer.*' My heart warms to a man who after a long hot day in London stepped off the train at Kingscote and was driven by his chauffeur to the lower lakes so that he could swim home. The legend of the naming of the Bluebell Railway, running from Kingscote to Sheffield Park, stems from the bluebells that have naturalised after being thrown out of the train window by Robinson.

Hardy trees from around the world were raised, the majority from seed: 120,000 in 1889/90 followed two years later by 4,000 cedars of Lebanon, Riga or Scots pine, Austrian or black pine and the closely related but more ornamental Corsican pine. Topiary is a popular feature of great and cottage gardens alike, but not for Robinson who viewed them along with clipped yew or box hedging as one of the 'evils of disfiguring trees'. The manor house and its walls have the patina of centuries and appear to grow from their own site; he used fencing crafted from the woodland. He built a pergola with brick pillars, in order to train climbers over the old stables. Many new climbing plants were being introduced from China and Japan, so in 1900 he enclosed the West Garden with a pergola made of oak and larch which he planted with wisterias, roses and vines. Two years later he paved it with old stones from Christ's Hospital in London. This joined the summerhouse he had commissioned from the architect Ernest George in 1900. He completed the paths with old London paving made of York stone. The textured surface and crimped edges of the light magenta-red blooms *Clematis* 'Ernest Markham' are named for his talented head gardener.

Hardy water lilies in an array of colours were introduced by Bory Latour-Marliac from the 1880s until his death in 1911. Robinson visited the nursery in Lot-et-Garonne. Through his columns and horticultural network he became Latour-Marliac's greatest ambassador, commissioned articles for *The Garden* and grew dozens of varieties. The key namesake dating back to 1895 with Latour-Marliac is *Nymphaea* 'Robinsoni' '*excessively*

Gravetye, flower garden, manor and landscape
The High Weald as scene and setting, framed in flowers Gravetye Manor revels in spectacular scenery. (Gravetye Manor)

Edging of *Nigella, Escholtzia* and thymes, Kitchen Garden, Gravetye Manor
The Kitchen Garden fulfils its role of beauty and utility, it is now seriously productive and visually pleasing. Here the edges of the beds in June with Love-in-a-Mist, Californian poppies and thyme cascading colourfully down to the path.

Clematis at entrance with view to lakes, Gravetye Manor
Pergolas are a staple of the arts and crafts garden. Robinson built one with massive stone pillars as the entrance from the east side of the house to be covered with wisteria, and one from the stable to the kitchen using bamboos and oak supports in 1897.

hardy and at the time when I will send it to you, it will not need shelter in order to assist its growth; it is a robust gaillard (fellow), which does not need to be treated like an invalid and which asks only for open air.' It can be seen in the pool in the Flower Garden. Latour-Marliac also sent boxes of the local Agen prunes and cases of wine. Tragically in 1909 Robinson fell from a stile on his way to West Hoathly church, sustaining a back injury that left him wheelchair bound. In 1922 he commissioned a half-track Citroën car so that he could take guests around – he died here in 1935, three years off his century. Local children came every year to dance on his birthday, they sang 'All things bright and beautiful' at his funeral.

A gentle climb up through heaths, across the Croquet Lawn, brings the gates to the Kitchen Garden into view. Its walls were built to Robinson's specification in 1898, elliptical in shape on a south-facing slope (no frost pockets as freezing air gently rolls out of the north-facing ironwork gates). In 1905 Robinson translated Vilmorin-Andrieux's *The Vegetable Garden* – which takes the reader from artichoke to yam (Chinese) with notes on shelling peas and netting melons. Of all the restorations that make Gravetye extraordinary, today's head gardener, Tom Coward, has created a space of productive beauty that surely would have earned praise from the man himself. Below the West Garden is the newly planted orchard which in turn leads to Robinson's 1920s Peach House, restored in 2012, which is in crop from June into August with the peaches 'Amsden', followed by 'Peregrine' and 'Redhaven'. There are also glasshouses, cold frames and a polytunnel as backup for all parts of the gardens, in the case of the latter for forcing early potatoes. Beauty and utility as every herb, vegetable, fruit or edible flower finds its way into the cooking and food preparation.

Robinson was a good friend to Walter Ingwersen, providing him with land to found the eponymous Ingwersen's nursery in 1927 (also supported by Borde Hill). Later, his son Will Ingwersen wrote: '*It was not until, as a very young man, I had given him evidence of my genuine love of plants and interest in garden design, that he admitted me to his friendship . . . until his death in 1935 at the advanced age of 97, he was our very good friend and neighbour.*' Gravetye went into Sleeping Beauty mode for 30 years until hotelier and restaurateur Peter Herbert bought it. He extended the north-east wing, restored the gardens and refilled the lower lake. It was a country house hotel run to the highest standards with no clocks on view so guests could relax timelessly. His tenure tied in with my first experiences. My heart broke when I saw the gardens in February 2010 at their lowest ebb, but that very year Jeremy and Elizabeth Hosking stepped in and saved it once more. In 2018 they hosted the 60th anniversary of its incarnation as a hotel with a new glass-fronted dining room where the gardens can be imbibed from inside and out. The garden is also brought into the house in the form of magnificent fresh flower arrangements. In the entrance vestibule are wellington boots in every size so that overnight guests can walk the lakes or go to West Hoathly to visit the church and Priest House. An alternative route takes the walker to the wider landscape – Gravetye lies on the High Weald Landscape Trail. Head west for Wakehurst and Borde Hill, or east for Standen.

Peach House, Gravetye Manor
Beauty and utility, William Robinson's Peach House which was restored in 2012. The peaches are in crop from June into August by growing the varieties Amsden followed by Peregrine, and Redhaven. The back wall is used for figs. There is a good chance of getting crops twice a year when they are under glass. (Gravetye Manor)

(Following pages) Flower garden, Gravetye Manor
Hardy herbaceous flowers fill the flower garden in June. The quadripartite grass was also planted up in Robinson's time. The gardens climb to tall trees and a more naturalistic landscape

STANDEN

'beauty, imagination and order' [1]

Just south of East Grinstead, a narrow, winding road descends from Hill Top Shaw, bounded by sandstone on one side and breathtaking views over Ashdown Forest on the other, and leads ultimately to Goose Green. James and Margaret Beale made the same journey in their quest for a rural retreat for themselves and their seven children, with good train connections to London. That vital link had arrived at East Grinstead in 1884. The Beale family were successful lawyers in Birmingham who had made a fortune handling the Midland Railway, finally negotiating its London terminus at that extraordinary monument to Victorian architecture – St Pancras station. Newly married, James and Margaret had moved to London in 1870, he Unitarian and she Non-Conformist, their tastes were for excellence, comfort and good workmanship, not lavish (within their circle of friends many favoured Morris & Co.).

In 1890 the Beales bought Great Hollybush, Standen and Stone farms. The 12-acre site of Standen, with its remnants of an 18th-century garden and orchard, was selected for their new house. Wishing to establish the gardens they commissioned G.B. Simpson, a London landscape gardener. Simpson's High Victorian plans for geometrical layouts of flowerbeds and shrubs were exactly what Williams Morris and Robinson loathed. Simpson had already levelled the land when their newly commissioned architect, Philip Webb made his first visit on 11 April 1891. Webb immediately expressed horticultural 'Anti-Scrape' views, suggesting siting the house further up the slope so that it nestled into the landscape, thus saving an ancient yew. His design meant that its servants' wing overlooked Goose Green without impingement over the traditional green. The final approach to the front courtyard at Standen is almost out of sight under an arch between Great Hollybush Farm and the servants' wing. Webb included a terrace walk to the south of the house, bordered by a simple trellis. The gardens evolved with Margaret Beale. Fortunately she started a garden diary which provides evidence of its development, her tastes and ability to adapt. She liked Simpson's exotic flowers in planting, balanced with arts and crafts principles of creating a landscape in harmony with the buildings. In many respects her views align more with Gertrude Jekyll than William Robinson.

The house was built of local materials which had involved opening a new quarry west of the conservatory for the stone; local bricks, clay tiles and timber were all incorporated. The approach drive was created alongside the old quarry. Margaret Beale's Japanese acers are now fine mature specimens which at the time would have been of great interest.

In 1910 the Beales approached J. Cheal & Sons of Crawley, as had the Astors at Hever Castle. Apart from their reputation as nurserymen, gardeners and landscape designers, Joseph Cheal was a Quaker which might have been another factor in their choice. He ran the nursery with his brothers Alexander and John. In 1906 they named an apple which had been found in a Crawley cottage garden in 1870 'Crawley Beauty'. Gertrude Jekyll included photographs of their superb topiary in her books, which would not have pleased Robinson.

This was a family house and garden: a small red trolley was kept for children wishing to speed down the sloping lawn; a rose garden created for colour and scent was later replanted with bamboo, and there were also bathing pools and a grove of rhododendron. The Beale grandchildren used the soft grass to career down whilst learning to bicycle when they were not playing at being noble savages hunting with bamboo sticks – neither of these activities is encouraged by the National Trust today! A court for tennis was created on what is now known as the Croquet Lawn.

On the death of Helen Beale, Standen was passed by bequest to the National Trust in 1972. The large kitchen garden has been beautifully restored with some original espalier fruit trees providing gnarled divisions there is also an orchard below the Croquet Lawn. The interior, with its wonder of stylised leaves and flowers, pays homage to William Morris. The main rooms are south-facing and at one with the landscape. Standen demonstrates Morris' sentiment: '*I must have unmistakeable suggestions of gardens and fields, and strange trees, boughs and tendrils, or I can't do with your pattern.*'

1. Full quote by William Morris: '*Ornamental pattern work, to be raised above the contempt of reasonable men, must possess three qualities: beauty, imagination and order.*' Cited in *The Collected Works of William Morris: With introductions by his daughter May Morris . . .* (Cambridge University Press, 2012), p. 179.

Red valerian, House, Standen

Philip Webb sited Standen's house at the top of the slope. Here the lawns and south terrace viewed through the flowering red and white valerian (*Centranthus*) which are extremely hardy and ideal for naturalising. There is a ha-ha at the bottom of the lawn so that the views are uninterrupted. (Patrick Wang/Shutterstock.com)

South terrace, Standen

The south terrace running parallel with the conservatory was started in October 1893. Apart from warmth and shelter it enjoys spectacular views to Hollybush Wood. Now as then these can be enjoyed inside and out. The conservatory was filled with exotics such as palms. Margaret Beale kept her plant catalogues for alpines – nearby Ingwersen's Nursery being specialists – Japanese trees and shrubs, and other flower sources. (Mike Longhurst/Shutterstock.com)

Plane tree base, Goose Green, Standen
Goose Green with a venerable plane tree overlooked by 15th-century Great Hollybush Farm, brick buildings and an oak frame barn. The modest approach to Standen. (Caroline Holmes, reproduced by kind permission of the National Trust)

Quarry Garden Fernery steps Standen
William Robinson and Gertrude Jekyll advised using steps rather than slopes in the garden. Standen has two approaches to the Fernery, one that leads to a point where you can admire its green shady depths from above, whilst the other takes you through it. (Caroline Holmes, reproduced by kind permission of the National Trust)

Quarry Garden, *Dicksonia*, Standen
Tree ferns (*Dicksonia*) originating in Australasia add height and drama to fern gardens, their trunk is formed from the build-up of the remains of earlier decaying growth. In 1896 the new quarry was planted up as a rock garden with heathers and conifers, at its height it had its own dedicated gardener looking after the 535 plants added by Backhouse & Sons from York (praised by William Robinson and Gertrude Jekyll). During the winter of 2019–20 it underwent restoration. (Caroline Holmes, reproduced by kind permission of the National Trust)

House entrance Standen
The north-facing entrance, is hidden away from Goose Green, all the main rooms are south-facing to enjoy both the warmth and spectacular views. (Caroline Holmes, reproduced by kind permission of the National Trust)

SMALLHYTHE FARM

the sweet smell of success

Much has been written about how the Great Storm of 1987 devastated the High Weald landscape. In 1287 the southern coast of England was redrawn by ferocious storms. Although half of the castle fell into the sea, Hastings survived but other High Weald ports such as Winchelsea, Rye and Tenterden effectively disappeared. In 1449 Tenterden joined the Cinque Ports confederation as a 'limb' of Rye, to supply ships and men against French raiders; St Mildred's Church served as a beacon for ships. From the later Middle Ages nearby Smallhythe had an important shipbuilding industry, and the setting of Smallhythe Farm still has the air of a coastal property.

Twentieth-century Smallhythe Farm provided a rural backcloth for artistic creators of stagecraft, literature and needlework hosted by Ellen Terry. In January 1941 Virginia Woolf encapsulated the great actress: '*Something of Ellen Terry it seems overflowed every part and remained uninacted. Shakespeare could not fit her; not Ibsen; nor Shaw. The stage could not hold her; nor the nursery. But there is after all, a greater dramatist than Shakespeare, Ibsen, or Shaw. There is Nature. Hers is so vast a stage, and so innumerable a company of actors, that for the most part she fobs them off with a tag or two. They come on and they go off without breaking the ranks. But now and again Nature creates a new part, an original part . . . That was Ellen Terry's fate – to act a new part. And thus while other actors are remembered because they were Hamlet, Phedre, or Cleopatra, Ellen Terry is remembered because she was Ellen Terry.*'

When Ellen Terry died in 1928 Smallhythe Place had been her home for more than 30 years. Her daughter Edith 'Edy' Ailsa Geraldine Craig determined to make it both a shrine to and a showcase of Ellen's remarkable theatrical collections and connections. A year later Edy opened two rooms in the house. The entrance fee was 6d (3 pence), the Barn in the gardens was converted into a theatre and in 1931 the Barn Theatre Society was founded, and productions continue today. She hoped in vain that a memorial trust fund would raise £15,000 to buy the house from her, to maintain and run it as a museum. A decade of dramatic difficulties followed until Vita Sackville-West suggested Edy should donate Smallhythe to the then tiny National Trust, putting her in touch with its Secretary James Lees-Milne. In 1939 the Trust accepted the house and gardens.

Front door overcanopied, Smallhythe Farm
Smallhythe Farm was built close to the road with a typical cottage up the garden path approach off the village street. The effect is a Shakespearean welcome: '*For you there's rosemary and rue: these keep Seeming and savour all the winter long: Grace and remembrance be to you both*' (*The Winter's Tale*, Act IV, Scene IV). (Caroline Holmes, reproduced by kind permission of the National Trust)

There are two approaches today: through the gate off the village street which leads to the front door and entry into a treasure trove of theatrical memorabilia. Alternatively, from the church of St John the Baptist passing the joint graves of Edy's dearest friends and partners, Clare 'Tony' Atwood and Christobel 'Christopher' St John, and a memorial stone to Edy raised in 2012. Wild flowers have naturalised the path towards Christopher's south-facing writing/smoking hut and a herb garden, the Priest's House (see next entry) can be glimpsed from the well beyond, and over orchard trees are the chimneys of Smallhythe Place. A rose arbour runs from a gazebo

St John the Baptist Church and Priest's House, Smallhythe Place
The door to St John the Baptist Church bathed in sunlight, beyond is The Priest's House where Ellen Terry's daughter Edy Craig lived. They line the village street that leads to Smallhythe Place and the remnants of the River Rother.

by the road almost to the boundary, this was probably originally made from hop poles, one of several that Ellen had. Stepping under it to the croquet lawn leads along a path and through a hedge planted to save Edy from ever seeing Ellen's much disliked third husband, James Carew. It was Carew who probably planted the orchard, which is balanced by a geometric nuttery of cobnuts and filberts. Lawns sweep up to the buildings all set in beds of drifting flowers. Positioned between the old toll gates and the house is the newly revamped rose garden.

Ellen Terry's career meant she was mostly London-based. There is a good archive of her correspondence with her employees, notably Edward Ladd, her gardener. Sweet peas were her favourite flowers; she loved fragrant flowers such as lilies, night-scented stocks and roses, growing beneath the house windows. There was still some rationing in 1918 and vegetables were expensive in London, so Ladd was charged with ensuring a supply of peas, beans, leeks, onions, potatoes and spinach with herbs such as chives, mint, parsley, sage, tarragon and thyme. Water takes two forms, the first pond probably a former tank is tucked in amongst trees on the lawn, the second was large enough for a rowing boat on which Ellen liked to relax; it acts as a mirror to the house and barn.

'Choosing' is arguably the most famous portrait of Ellen Terry by her first husband G.F. Watts. Its floral motifs and symbolism encapsulate her enigma. Despite the unhappy marriage it was her favourite portrait. Her lovely young face is drawn to the blousy scentless camellia and yet in her hand close to her heart are sweet violets. A visit to Smallhythe is to discover what else was close to her heart.

Smallhythe Farm from track by former River Rother
The gardens of Smallhythe Place are enclosed by thick country hedges, the hawthorn hedge and trees here line the path that runs along the now defunct River Rother which in its past made Smallhythe a thriving place

THE PRIEST'S HOUSE, SMALLHYTHE

a spirit of pioneering power

Priest's House, Smallhythe
The Priest's House encapsulates Wealden architecture rising from a typical cottage garden of flowers. The back garden would have been dedicated to food crops, hens and possibly a pig.

Sir Robert Brygantyne is believed to have lived and worked here before he became Clerk of Ships to Henry VII and Henry VIII from 1495 to 1523, and supervised, among other ships, the design and construction of the *Mary Rose*. It is shown as a vicarage on 1870 and 1898 maps. From 1899 until her death it was home to Edy Craig; it is now private property. The Priest's House played a key part in the later Ellen Terry story and legacy thanks to Edy. A trailblazing theatre director, producer, costumier and actress, Edy produced or directed hundreds of plays and pageants. She founded the avant garde Pioneer Players which staged plays dealing with social, political and feminist issues. Educated as a child by the pioneering suffragist, Elizabeth Malleson, she was committed to the fight for the equality for women not least for women's suffrage. She lived here in a 'ménage à trois' with her female partners Christopher St John (Christobel Marshall) a writer and Tony (Clare) Atwood an artist. Suffrage activists were welcomed along with other women visitors with same-sex relationships, notably the writers Virginia Woolf and Vita Sackville-West. Like her mother she lived and worked in London as well as Smallhythe. She is described as having a 'ferocious work ethic'.

While Edy's brother, Edward Gordon Craig, has long enjoyed a reputation as a theatrical innovator Edy's work has been largely overlooked. In 1896 George Bernard Shaw observed: '*Gordon Craig has made himself the most famous producer in Europe by dint of never producing anything, while Edith Craig remains the most obscure by dint of producing everything.*'

The environment at Smallhythe was creative and productive. While with Edy, Christopher was a working journalist who found time to write two novels and collaborated on at least 25 plays. Christopher worked very closely with Ellen Terry – calling herself Terry's 'literary henchman' – editing Terry's autobiography, memoirs, and correspondence with George Bernard Shaw. In 1916, to counteract Christopher's emotionally volatile personality and sometimes violent temperament, Edy invited the even-tempered artist Clare 'Tony' Atwood (1866–1962) to join their lives. Vita Sackville-West described Tony as having '*a gift for drollery . . . she restores the balance*'. Tony fitted seamlessly into the hectic, creative life at Smallhythe, working on her own art and helping Edy and Chris with productions for the Pioneer Players and Barn Theatre Society.

THE PRIEST HOUSE, WEST HOATHLY

Hope smiles from the threshold of the year to come . . .
The Foresters, Alfred, Lord Tennyson

West Hoathly stands 600 feet (183 metres) above sea level in the Sussex High Weald. It makes a delightful walk for those interested in English villages. There are nine centuries of styles and a multitude of uses for local materials, the Manor House to which the Priest House belonged being a fine example of an ironmaster's status home. It was originally known as The Great House when it was built in 1627 for Katharine Infield, widow of Richard (see Gravetye), at a reputed cost of £1,000.

The cottage garden encapsulates the notions of an English garden. In the early years of their friendship Gertrude Jekyll took the young Edwin Lutyens out in her donkey cart to learn about the West Surrey vernacular. Her neighbour and friend Helen Allingham painted many cottage scenes, mostly idealised, including the Priest House entitled *Old Sussex Cottage*. Initially it was a 15th-century open hall with a living room and upper chamber in the north end, to the south a service end with buttery, pantry and solar. It may have been an administrative centre for the Cluniac Priory of St Pancras in Lewes; *c.*1600 the upper floor and chimney were inserted. As part of the Manor it was sold in 1731 to Robert Bostock of Otford, Kent whose family held the land for 150 years. The dilapidated Priest House was lived in by labourers working for a tenant farmer in the Manor House. Finally, in 1905 John Godwin King of Stonelands, West Hoathly, took ownership, realised the historic value of the Priest House, and undertook a major restoration in the spirit of Morris and Webb. In 1908 Godwin King opened it to the public with a museum in the north end to display his collection of locally gathered artefacts, with a resident Custodian. In 1935 it was presented to the Sussex Archaeological Society. Horsham stone tiled and timber framed this building, filled with domestic treasures within a gentle cottage garden, would delight any arts and crafts enthusiast.

Espaliered apples, Priest House
Espaliered apples maximise the ground available. Note the handcrafted windows and low-hanging eaves. Small windows kept the warmth in and plague out.

(Following pages)
Outer hedge and topiary, Priest House
A glimpse of The Priest House on West Hoathly's village street which curves down to Luctons. Today the warden divides his time between manning the entrance and tending the gardens. In Allingham's painting the hedge is altogether unclipped and looks like hawthorn.

LUCTONS

diligence and reward

By comparison with the Priest House, Luctons is a mere baby of about 150 years old. Both doctors, Ingrid and Hans Sethi moved into the house on North Lane (formerly The Straight) in July 1984. They are the fourth owners, with photographic records of the gardens in their earlier incarnations. One of their first tasks was to plant vines in the greenhouse, including the delicious 'Black Hamburg' and 'Muscat D'Alexandria'. A framed selection of eight photographs titled 'Luctons 1987' starts in March with daffodils clustered round tree trunks; in April snakeshead fritillaries add purple to the yellow and white palette in the orchard – they continue to thrive preceded by crocus; the sun beats down in July as the full extent of the gardens is framed by two venerable Scots pines. A month later in a view back to the house, the double herbaceous borders are in Jekyllian splendour. How the scene changes in October 1987: both pines are prone, as is a mature tulip tree (*Liriodendron*). By November clearance work is underway with a new canvas awaiting ideas.

Ingrid and Hans have a monthly calendar of work, they are great followers of Jekyll and, especially Hans, indefatigable propagators by seed and cuttings. They have installed water throughout, and there is an excellent system of compost bins, to which they add mushroom compost and chicken manure. A formal garden of clipped box hedges where topiary peacocks are underplanted with salvias, frames the front of the house. This seems deceptively small in comparison with the extensive gardens behind, which are dominated by a long central vista with surprises to left and right along the walkway. There is an extensive terrace running behind the house leading to the garage, which is disguised by a garden loggia. The upper grass area from the terrace to the lower gardens formerly framed by the pines, retains its shape courtesy of a brick wall underplanted with pelargoniums and hardy geraniums. The walls terminate with stone urns of flowers, along its length are immaculately clipped cone sweet bays. Beyond there is a magnificent tree with tree house. The 150-foot (46 metres) double herbaceous borders designed to peak in August, can be viewed from the front, middle and back, and are designed to do so. Twenty barrows of material are lifted out in September and October, successes are noted and cuttings taken, especially of the tender *Salvias* 'Cherry Lips', 'Amethyst Lips' and 'Love and Wishes', which are much used in the front of the borders. Spring planting takes place in April. Behind the colourful borders is a green enclosed garden with a lily pond at its centre. A clipped yew hedge with arched iron gateway terminates the garden whilst leaving the eye free to roam into the fields.

Lying on the left, behind the borders, is the orchard, productive in fruit (such as apple 'Spartan' and apricot 'Tom Cot'). Roses including 'Gertrude Jekyll' grow amongst the trees, a possible 6,000 orchids flower alongside the fritillaries through the grass, which is left uncut until mid-August. There is an excellent kitchen garden with frames for peas (sweet and green) and beans, an asparagus bed, pumpkins of Cinderella proportions, and lots of salads. The three greenhouses are put to full use, once rooted *Salvia* plants also find a place in the front garden. Scents waft and lurk throughout from the chocolate cosmos in the front garden, scented white jasmine on the back of the house, and the winter wonder, *Daphne bholua* 'Jacqueline Postill'. This is a garden for horticultural pleasure and family enjoyment.

Apple trees, borders to house, Luctons
A view back to the house over the double 150ft borders, the asters and sedums catching the sunlight, lots of architectural plant spires with seeds forming for the following year. The apple trees are laden with the bright red variety 'Spartan', bred in Canada it is very hardy, crisp and crunchy.

(Following pages) Grass terrace view, topiary bay, borders, Luctons
The simple clipped cone of bay starts a series of visual topiary stepping-stones that invite exploration of the gardens. The boundaries within and without offer a contrast of tree and shrub outlines.

***Populus nigra* between orchard and gardens, Goddards Green**
A black poplar (*Populus nigra*) silhouetted against the blue autumn skies, balanced by the ever reliable free-flowering Cosmos. Cosmos thrive in hot conditions, their crinkly open flowers cupping the sun.

GODDARDS GREEN

a seasoned landscape

Approaching from the south the road divides at a small triangular garden accommodating the Cranbrook War Memorial, a field gun and a plough share, the enclosure planted with drifts of rosemary for remembrance. An orchard of fruit and nuts lies to the left before the drive that leads to Goddards Green. John and Linde Wotton bought the 500-year-old property in 1992 from Colonel Pearson, a keen horseman who as he became blind planted roses for their scent. The wall and hedge structure of the gardens had changed little from the 1920s when it was owned by the poet, prolific author and playwright, thrice-married E. Temple Thurston. An insight into his character that could be applied to garden design, appears in a letter he sent from Goddards Green in 1922: '*Of course, every author has his terminological hatreds . . . Words and phrases certainly have the effect of "turning me up" like an unpleasant scent or an ugly clash of colours . . .*' Little did Thurston realise that it is said that Nathaniel Courthope was born here, Nathaniel of nutmeg fame whose true and incredible adventures in the 17th century as a spice trader changed the course of history.

There is a seasonal stream that runs through the garden to a naturalistic pond. It is filled by run-off from the adjoining fields, a short stream providing the outflow. Water takes two further very different forms: the original water garden and new fountain garden by the south façade of the house, and the swimming pool tucked away by the barns to the north of the house. Although this is a garden of rooms, the areas between allow its enjoyment to its full extent, the magnolia is given full rein and the lawn is used for croquet. Gradually over the space of 10 to 15 years, garden designers Ed Flint who had worked at Great Dixter, and his wife Nicky created the Fountain Garden, Swimming Pool Garden, Fern Garden and introduced new borders. The Wottons later worked out further ideas with Head Gardener Jacqui Harris-Jones, whose design contribution has enriched the plantings – there is a banana tucked in a border – and enlisted the help of their son, Tom, as Under Gardener.

The long garden wall to the rear of the house provides a sheltered south-west aspect that is good for sun-loving plants, and an almost year-round loggia. As you walk around the gardens you appreciate how carefully areas – small and large – have been chosen for their plants. One of the first was a group of silver birches at the south end of the front court, both concealing and revealing glimpses to the former cherry orchard beyond. Borders were widened to include the attractively thorny stemmed *Rubus cockburnianus* underplanted with bulbs.

The Old Cherry Orchard has had a reincarnation as an arboretum centred on a *Liquidambar styraciflua* 'Worplesdon', this and most of the other trees were obtained from the appropriately named Cherry Tree Arboretum in Staffordshire, now sadly closed. Other trees include the tulip tree (*Liriodendron sinensis*), hop hornbeam (*Ostrya carpinifolia*), and the Chinese tupelo (*Nyssa sinensis*). The grass, planted as a meadow with a single cut in August, is dotted with guest beehives. John describes their methods as 'pragmatically eco' in that they use their own compost and wood chippings. A leisurely approach to the house can be taken through the Fountain Garden into the Water Garden where yellow and white dominate the planting colours.

A narrow path along the house leads to the front court, passing the willow-leafed jessamine (*Cestrum parqui*), which often flowers three times a year whilst the tender fragrant yellow flowers wrapped in white silky hairs of *Edgeworthia* waft their scent from winter into early spring (the latter was used in Japan to make quality paper for currency). The sheltered Hot Border runs between the front of the house and the silver birches with *Magnolia* 'Caerhays Belle', under the birches is *Clethra barbinervis.* Opposite, a long border includes *Camellia*, black poplar (*Populus nigra*), and the smoke bush (*Cotinus*).

Beyond is a former paddock, now the 'new' orchard with apples such as 'Beauty of Kent' and 'Brownlees Russet', the plum Marjorie's Seedling (excellent for jam and crumbles) and medlar 'Large Dutch'. Pruning starts in the New Year, excluding the plums which should never be cut in winter. Tucked in the shade by the 'new' orchard is a Fern Garden. Leaving the house frontage and along the drive the wall is dotted with purple-leafed beech, the scented September-flowering *Elaeagnus pungens maculata*, *Indigofera* and hydrangeas. At the end of the house the borders deepen, filled with floriferous herbaceous perennials and grasses. Finally, the Swimming Pool Garden majors on blue, grey and silver; it is sheltered and contemporary alongside the magnificent kitchen garden. Delightful scents and clever colour combinations greet the visitor here, with lots of ideas for tree planting.

***Ostrya carpinifolia* (hop hornbeam), arboretum, Goddards Green**
The old cherry orchard has been cleared and replanted with a fine collection of trees. Here in the sunlight it is clear why *Ostrya carpinifolia* is known as the hop hornbeam.

***Eleagnus pungens maculata*, September scent, Goddards Green**
The flowers of *Elaeagnus pungens maculata* are tiny but they exude a delightful fragrance in September.

UPPER PRYORS

game, set and match

This traditional timber-framed Tudor house nestled in shrubs and plants intrigues with its gardens falling away to a tree canopy of oaks and silver birches. Marilyn and Stuart Smith bought the property with 10 acres in 1981, when it was surrounded by a cottage-style garden overlooking a field of Christmas trees which proved to be too much like hard work. The 1987 storms cleared many of their mature trees and lifted the roofs from the outhouses. They now provide useful storage and hide the car parking. Two years later they extended the house and commissioned Ryl Nowell to redesign the gardens, with many of the original plants being recycled. They found mixing their clay soil with mushroom compost proved an excellent growing medium.

After they lost the trees to the west, a series of water gardens was excavated. After initial leaks and early teething problems the water is now artificially circulated to ensure regular flow. The course of the water is outlined by sinuous borders varying in shape and depth, the final pond, being the most naturalistic, makes an ideal visual transition to the countryside beyond. In 1991 they undertook their first of many openings for the National Garden Scheme.

Stuart says his love of gardening came from his policeman father and the cottage garden influence was thanks to his mother. In their retirement they lived in the Smith's cottage opposite. Upper Pryors' house façade is in the arts and crafts tradition: a grassy approach with a narrow cobbled path running along its length, a slim flower bed between, two silver birches add a contemporary note. The main front door is now in the east side with modern brick and stone sets as paving, the outlook is across a striped lawn to the water gardens and the High Weald beyond, an arch draws the view to the south. In contrast, the architecture at the back of the house is a mixture, the main section is tile-hung from which a 1920s sunroom and their own extension create a sheltered enclosure. Wisteria engulfs the corner walls, and a white climbing rose complements the tiles. The local stone and brick paving is well weathered, sporting square stone planters. Despite the English backdrop the terrace has a distinctly Mediterranean-meets-Jekyll air with self-sown *Escholtzia* dotted around and rockery plants cascading off its edges in summer. Initially the Smiths worked on pink to orange colour schemes for the terraces that would enhance the setting. One favourite being peonies in hues ranging from carmine through deep rose to palest pink amongst which many silver-leaved plants are interspersed. The paving is softened with naturalised *Erigeron* from which visitors step onto well-tended lawn enclosed by borders. This looks onto an extensive field that doubles as a football pitch, outlined with trees and shrubs, which was formerly covered by the long-gone Christmas trees. Skirting the grass amongst imaginative plantings a golden-fruited crab apple is outlined by a long drift of golden variegated euonymus. A curved brick path leads back towards the house between two borders. Whilst agapanthus routinely fail, irises and roses have thrived. Running through and gently dividing the garden areas are undulating deciduous and evergreen, gold and purple-leafed hedges such as hawthorn mixed with viburnum and ivy and yew.

Within protective hedges are productive beds not least with a wealth of leeks and globe artichokes as well as rhubarb and a fruit cage. The all-important compost bins are in great shape. Elsewhere the curvilinear greenhouse was already in place and managed to survive the 1987 storm. Since 1994 Nick Lynch has gardened here one day a week. Through the apple orchard from which they make their own apple juice, further hedges enclose and protect the swimming pool which in turn leads to the tennis court. Serious tennis contests are organised for friends from all parts of the world who enjoy some fierce competition and great hospitality. Sentiments which make this garden vibrant and engaging.

CHAPTER FOUR

Shaking off the historical shackles

Some years ago the excellent, talented and unashamedly opinionated garden designer, John Brookes told me that we lived in the tyranny of Jekylldom. How delighted he would have been with Chris and Robin Hutt at Fairlight End who had read and tried to follow Miss Jekyll's inspiration but came to form the opinion that comfort can be married with excellence in terms of horticultural upkeep. The contemporary gardens at Fairlight End looking out to sea, and Wych Warren on the edge of Ashdown Forest, continue to develop in the individual style of their creators. At Great Dixter, Christopher Lloyd was admonished for replanting his mother's traditional rose garden with exotic plantings – exuberantly colourful and tropical – revolutionary in the 1990s. Lloyd's legacy is a living canvas that Fergus Garrett continues to move forward whilst the constant of original creator Nathaniel Lloyd's topiary holds the contemporary in timeless focus. Charlotte Molesworth at Balmoral Cottage has taken topiary to new heights in every sense, clipped creatures watch your every step whilst the notion of Platonic solids extends from the classical towards Edward Scissorhands and the Turner Prize in its scope.

The horticultural history of Leonardslee is rooted with the other great 19th and early 20th-century creations of Borde Hill, Nymans and, bearing in mind their Loder roots, unsurprisingly Wakehurst Place and High Beeches. Its 21st-century landscape was exhausted and neglected although much of its arboreal and rocky bones had survived. A major overhaul is creating a dynamic future referencing its past. Colwood House and Wych Warren with Twyford House (see pages 153–155) started life as large country houses surrounded by extensive gardens. Now, the respective owners live in a substantial wing with gardens that enhance this new status and erase the idea that they were once part of a larger whole. Balmoral Cottage and Colwood House offer inspirational ideas for recycling/upscaling that counter modern waste whilst providing a haven for wildlife. On a smaller scale the RHS Partner Garden Merriments is undergoing renewal with a nursery to match.

Tree ferns in lower lawn, Wych Warren
There is almost a primeval feel to the lower lawn whose margins offer views beyond of Ashdown Forest and an Australian accent with the tree ferns which are settling in well.

FAIRLIGHT END

'What is this life if, full of care,
We have not time to stand and stare.
'Leisure' by W.H. Davies

This two-acre garden and its additional one acre overlook the end of the High Weald between Hastings and Winchelsea, the views across the Channel in marked contrast with the usual wooded panoramas. Trees, venerable or new, clipped or natural, are still an integral part. Chris and Robin Hutt bought the property in 2003 and moved in for Christmas 2004, ready to start the garden the following year whilst restoring the house, which is a constant backdrop. Dating back to *c.*1720 when it was a one-up, one-down cottage, two chimneys were a modern addition in 1760. In 1920 the owners extended, with further construction in 1970 that encompassed the two chimneys. The local sandstone has been put to good effect in terracing and walls.

They rapidly realised that the price for the fabulous views was fierce blasts of wind, and staking all their plants was not an option. The North American coyote willow (*Salix exigua*) is one tree that copes with exposure to the wind, helped by the fact that they pollard them to one metre. They decided to rethink the hard grind throughout January and February in terms of design and planting. They travelled to Germany and Belgium seeking inspiration and plants. They have become ruthless in deciding which plants stay or go – asters are now on the to-go list. The hybrid cockspur thorn (*Crataegeus* x *lavelleei* 'Carrierei') was bought as a mature specimen in Hamburg. In the 'outside' garden they dug a spring-fed pond in 2006, whose margins were planted with 1,000 plug plants of 37 varieties of bogside natives. By 2009 they had started to create various levels around the house. The final step was in 2016 when they rationalised their front drive to increase both security and winter interest with fragrance, the glorious wafting scent of *Daphne bholua* 'Jacqueline Postill' being one of the main plantings. A local company manufacturing in Battle (Paneltech Systems) devised a metal sheet of varying height that cuts sinuously like a knife across the frontage, both echoing and contrasting with the architecture whilst providing year-round interest and structure for the planting. Metal straps with quoit-like rings pop up in the froth of one herbaceous border, supporting a sea of dahlias and Joe Pye weed (*Eupatorium maculatum*) with towers of *Verbena bonariensis* and bronze fennel. The Hutts have commissioned Paneltech Systems again for new developments in the 'outside' gardens.

At the outset their son was a keen cricketer, so an area of lawn was essential. From 2013 it evolved into a pattern of edged lawn and flowerbeds. This leads to a meadow with longer grass from March through to June until it is cut back in August/September. The idea has spread into a wet meadow area where thousands of orchids flower every year with the pin-cushion-like blue flowers of devil's-bit scabious and golden yellow daisy flowers of fleabane. This is cut twice, in May and October. The Hutts buy from High Weald Farms who collect and market seed from ancient hay meadows.

Six topiaries anchor the garden nearest the house, three each of deciduous and evergreen varieties such as hawthorn, holly and hornbeam clipped in Joan Miró style – abstract shapes that make you smile. The W.H. Davies quote above is perfectly answered with a pair of loungers framed by dwarf pine (*Pinus mugo* 'Mughus'), providing shelter and scent from which to lie and gaze. Chris enjoys hand-clipping, while Robin's artistic talents are manifest in both the garden and her watercolours in her studio. In 2018 the steps from the house were softened by plantings of *Stipa*, *Buddleja* and *Cotoneaster*, the gently descending steps disguised by the hazy effects of *Gaura*, *Stachys*, *Stipa calamagrostis*, and

Grasses through clipped shrubs, distant trees, Fairlight End
Walking through the gardens there is a sense of them reaching out for the countryside and coast beyond. The clipped hedges on the left contrasting with the haze of grasses and unsatisfactory asters (soon to go) as well as two Wealden oaks on the skyline.

(Previous pages)
Barn garden, oast house, Great Dixter
Traditional Kent architecture forms the backdrop, the original Dixter and the characteristic oast house built *c.*1890 with three kilns for drying the locally produced hops. The foreground is filled with an eclectic mixture of flowers in contrast to the clean plant-free paving – both designed by Christopher Lloyd.

Steps, *Eucalyptus*, Fairlight End
The gardens are on many levels, so the Hutts introduced steps from the house to Chris' studio to make the descent easier in the dark. The conditions are ideal for the *Eucalyptus*, and the weeping birch that frames the view, whilst sentinel foxgloves in flower line the route. (Courtesy Robin Hutt)

Kitchen garden, Fairlight End
More of a feast for the eyes than the stomach? Look a little closer through the haze of poppy seed heads, foxgloves and pineapple sage and there are some fine beans growing.

(Following pages) Sinuous shapes and textures, Fairlight End
The house sits at the top of the gardens near the road, the clipped trees and hedges glimpsed over a yew hedge reveal a garden of many parts. The dwarf-pine hedging shelters a place in which to relax and enjoy the spectacular views.

complemented by *Deschampsia*. Another area secluded by oaks and hollies has two Adirondack recliners from which to enjoy further contemplations.

The outer wall beyond the house is covered in roses such as 'Open Arms', an idea that came from Xa Tollemache of Helmingham Hall; they are David Austin roses. This leads to the kitchen and cutting garden, divided by raised beds with a productive greenhouse at its centre. Vegetables include an asparagus bed cutting for 10 weeks from mid-April through to the end of June. Brassicas, leeks and lettuces thrive with the raised bed edges softened by naturalised cushions of *Erigeron*. Soft fruits include raspberries and gooseberries. A new orchard has been planted below Chris' studio in the lower gardens, with a nuttery beyond. This is a garden packed with inspiration and pragmatism.

WYCH WARREN

'Re Robinson's volume . . . I have been on the point of coming to study your floral poetry for the last month' [1]

Coming through the Golf Club grounds it is apparent that this is an older designed landscape, mature redwoods and tall Scots Pine evidencing the former grounds of Ashdown Place (now flats). In spring and summer rare wild flowers spring up around the edges of the greens. Finally, dipping into Wych Warren's shady drive with rhododendrons, the house lies to your left and mixed woodland to your right. The house was built in 1929 partly to accommodate staff for the big house, and it was divided into two dwellings in the 1970s. Both properties inherited excellent stonework, not least in the terraces from which Wych Warren House's six acres are displayed, making full use of their Ashdown Forest frame.

When the medieval Ashdown Forest was divided into wards by Decree in 1693, its pale overtopped with holly and beech, Wych Cross was within the West Ward, much of which was a vast rabbit warren with two pillow mounds (medieval constructions for rabbit production) hence Wych Warren. The other part of the house is not apparent, the stone terraces feature a range of pots and seating that invite the eye to travel across the expanses of three main vistas: the extension of the terrace towards the greenhouse, woods and exotic plantings, or down towards informal grass and pools, while the central view undulates across grass and up into the Ashdown Forest beyond.

Colin King bought the property in 2000, moved away and returned with Mary Franck in 2014. They rapidly realised that theirs was also a good horticultural partnership – he is the manager of the woods and she the gardener. Mary grew up in a farming family in Hertfordshire, her parents were keen, productive gardeners with an infectious enthusiasm for both plants and birds – bird feeders hang from some of the small trees and shrubs. As a journalist Mary lived and travelled in Portugal and France, the possible genesis of her desire to grow tropical and semi-hardy plants. She is now co-owner of a local health club, connection with the immediate environment being equally important at home and at work.

In 2014 Colin started by removing many overgrown trees including pines and redwoods. He raised the canopies of those that remain, opening views across the gardens to the forest landscape. The woods are now a well-managed mixture that includes sweet chestnut, oak, beech and rowan with lichen-encrusted stumps and a healthy moss population, all confirming what a clean environment this is. A deer fence was installed to protect the new plantings and they find that road planings make an excellent surface for paths. Rather than restricting her palette, Mary references English gardens because of their range. She is an avid watcher of television gardening programmes, and garden visitor. Architectural Plants in Pulborough, West Sussex have been a good source, both for choice and for quality, of a good range of specimens.

The house terrace is edged with rosemary, the paths and borders below are essentially east-facing where peonies and lavender thrive. *Heuchera* in various hues, hardy geraniums and sedums weave into the mixture. This has been extended using a range of rocks, stones and paving to benefit from a southerly aspect. Arguably this is the most overtly designed area of the gardens, including a swimming pool, pots of cordylines, cannas and agapanthus, spires of sculptural red-hot pokers with colourful dahlias and the great palmate leaves of *Fatsia* which conspire to exude an exotic and tropical feel. Mary hopes to travel to Japan at some point, apart from some design influences Japanese acers are dotted throughout, their delicate leaves offering autumnal colour and contrast. This is a garden in constant intelligent development. Around the greenhouse Mary has been inspired by Nymans and Wakehurst Place to create a new low-profile heather garden with dwarf conifers that acts as a planted transition to the woodlands where pine and

1. Douglas W. Freshfield, Wych Cross Place, 12 June 1923 to William Robinson at Gravetye Manor

Hooped enclosure, banana, exotics, Wych Warren
The tunnel at once frames and opens out this part of the south-facing terrace, the bold red leaves of banana (*Musa*) appear to sail on a foam of *Salvia*. Beyond, figs, palms and cordyline contrast with the resolutely straight trunk of the pine. The use of strategically placed pots means the picture can be constantly changed.

holly self-seed. In the autumn wild fungi pop up as well.

The plan had been to punctuate the great grass slope, home to rare wild flowers, with a series of ponds which is proving difficult to establish, although tree ferns are growing and drumstick primulas have naturalised. The unmistakeable leaves and fruiting spathe of *Gunnera*, bulrushes and willows contribute to some success with the lower pond. In contrast, the ground at the bottom is boggy and naturalistic with ferns and mosses amongst stands of birches. This garden truly fosters a connection with its setting, breathe deeply and let its air fill you with delight.

GREAT DIXTER

'I take it as a challenge to combine every sort of colour effectively . . . Many plants in this garden are self-sown and they often provide me with excellent ideas.' Christopher Lloyd

Nathaniel Lloyd was the first to shake off established ideas. Sufficiently successful to retire at the age of 42, he and his wife Daisy wanted a historic property. Dilapidated Dixter, its grounds and farm buildings were acquired in May 1910 for £6,000. The next step was to find an architect, two were approached: Sir Ernest George, successful and well regarded in the field of designing substantial country houses with an established reputation for excellence, and, in total contrast, George's former pupil and protégé of Gertrude Jekyll, Edwin Lutyens. Rather than an Edwardian pile Lutyens designed Great Dixter, a restoration of the original hall house, dismantling and transporting The Old House at Home from Benenden and marrying it with his own building. Lloyd was delighted and inspired to study architecture; he authored *A History of English Brickwork* and *A History of the English House*. Like Robinson at Gravetye (see pages 74–75), Lutyens obtained former York stone London pavements for the paths – his hallmark approach to steps and terraces is still evident. Topiary became Lloyd's other passion. In 1925 he published *Garden Craftsmanship in Yew and Box*. Nathaniel died in 1933, Daisy lived until 1972. Born in 1921, their youngest son Christopher studied modern languages at King's College, Cambridge before World War II, then afterwards horticulture, including botany and science, at Wye College where he lectured from 1950 to 1954.

In 1954 Christopher returned to Great Dixter to open a nursery specialising in clematis and uncommon plants, including cuttings of rosemary 'Benenden Blue' given to him by Vita Sackville-West. Great Dixter became his canvas, the topiaries and his mother's orchid-filled meadow far from shackling him provided an arena in which to experiment. For those who beat a path to his door he was a generous host and great cook as well as being an inveterate garden visitor with a copious notebook to hand. He loathed plant labels and refused to use them. If asked he would not name a plant if the questioner did not have a notebook to hand; however, if they had a pen poised he delighted in sharing his knowledge. In 1963 he was invited to write regular weekly articles for *Country Life* which he continued for 42 years, he was also gardening correspondent for the *Guardian*, acknowledged as the best-informed, liveliest and most innovative gardening writer.

The first shackles released were the plantings in the Long Border. Christopher found a mixture of shrubs, herbaceous perennials, climbers, bulbs and the judicious placing of biennials and annuals orchestrated a more dynamic effect. His talent for passing on great garden ideas started in 1957 when he published *The Mixed Border*. Two further books followed in 1965 *Clematis* (with John Treasure) and *Trees and Shrubs for Small Gardens*. In 1970 he published *The Well-Tempered Garden* (as he was notably fiercely opinionated, he gained the sobriquet of the ill-tempered gardener). He and his mother worked together over the planting, design and opening of Great Dixter to the public until her death. In 1992 Fergus Garrett, who had worked for Christopher's great friend Beth Chatto, was appointed head gardener. Their first revolt was to remove that most English sensory stalwart, the Rose Garden, and replace it with the Exotic Garden, at its height from late summer into autumn. Their fruitful alliance lasted until Christopher's death in 2006, since when Fergus continues to experiment, allowing Great Dixter to evolve and share its secrets.

The Meadow Garden leads to the house. Daisy first introduced local orchids which are now established colonies. The regime is that cutting does not take place until the orchids have set seed and bird and invertebrate activity is diminished (effectively, after mid-August). The hay is cut and strewn to distribute its seeds, then collected, after which the grass is cut down to the ground in October/November. Some grass is also left uncut.

To the right lies the Barn Garden within which is the Sunk Garden with a pool at its centre; the surrounding

Meadow orchids, Great Dixter
Daisy Lloyd introduced local orchids such as the common spotted (*Dactylorhiza fuchsia*), Twayblade (*LIstera ovata*), early purple (Shakespeare's long purples, *Orchis mascula*) and green-winged (*Orchis morio*) which are now established colonies. As at Wakehurst there is an increasing appreciation that diverse meadows provide a valuable natural habitat.

Barn garden, white salvia, to house, Great Dixter
The tall white flowerheads of *Salvia nemorosa* brighten up a dull day and will shine by moonlight as well.

(Previous pages) Long Border to house in late May, Great Dixter
The Long Border is 135ft in length. The planting falls over former London York stone pavements. The floral exuberance contrasts with the meadow planting under the orchard on the left. Lutyens' chimneys and roofline complete the picture.

buildings and clipped yew hedging bring the haze of flowers into focus. Lutyens planted fig trees along the barn wall for their foliage effect. The *Buddleja* 'Dartmoor' is a haven for birds and butterflies. To the left are original yew topiaries with an ostentation of peacocks in a sea of grasses and wild flowers. The High Garden used to be where Christopher Lloyd experimented with massed plantings of herbaceous material (Fergus Garrett has subtly changed this), and beyond lies the bounteous kitchen garden. The backdrop of sweeping rooflines and chimneys is unchanging. Stepping down through the Orchard Garden, look back to the house and you will see a small window on the first floor which is in the former nursery. Lutyens put this at crawling level so that the youngest inhabitants could enjoy the view. The naturalised orchard lies ahead whilst the Long Border leads in colourful fashion towards the house and Lutyens' terraces. Stepping down once more leads you to the Exotic Garden through the building known as the Hovel and onto the Topiary Lawn. Ideally all the yew topiary is cut once a year in August, in reality it takes several months.

'*Never take the "I shan't see it" attitude. By exercising a little vision you will come to realise that the tree, which has a possible future, perhaps a great one, may be more important than ourself, nearing our end.*' Said by Christopher Lloyd, as Fergus Garrett continues to garden for the future.

Pond to oast house, Great Dixter
Often overlooked when visiting the glorious gardens of Great Dixter is the naturalised pond near the approach road.

Topiary, borders and house roofs, Great Dixter
Christopher Lloyd loved opera and designed borders for the Glyndebourne opera house. Here the house acts as lead to a symphony and chorus of formality and abundance.

Tropical garden, banana, Great Dixter

The Exotic Garden shelters the large leaves of the hardy Japanese banana (*Musa basjoo*) underplanted with cannas. A backdrop of red roses brightly reflect the area's past. The white-flowering *Escallonia bifida* is mobbed by butterflies.

BALMORAL COTTAGE

Sheer genius

This is a garden to put a smile on your face. The keen gardener will instantly recognise horticultural excellence and diligence but such prowess in no way detracts from the very magic of this special garden. Here you will find topiary to die for. There is something verging on acrobatic about the ensemble that reminds me of Chagall's biblical-meets-circus figures, there is even a topiary peacock alongside the outside loo. When Charlotte and Donald Molesworth were married, they asked for yew seedlings which they received in large numbers. They moved here in 1983 into a scene of dereliction and neglect. Nothing daunted, the yews have been nurtured to maturity, along with box, bay and hornbeam; all are now clipped into pleasing shape by Charlotte. A further unusual contender is a hawthorn cut into a chess piece – square base, trunk, circle, trunk with a ball on top but allowed to flower.

Tucked away, off Benenden's village green near the church, down a soldier brick path between two clipped box hedges topped with box balls on thin trunks, is Balmoral Cottage, on Collingwood 'Cherry' Ingram's former kitchen gardens for The Grange, which left a legacy of excellent soil. Ever observant, they have found seedlings from Ingram's fragrant crab apple, *Malus hupehensis*, whose white flowers follow pink buds producing red cherry-like fruits, also much loved and grown at Great Dixter. Mature Scots and Monterey pines (*Pinus sylvestris* and *P. radiata*) provide a general dark backdrop along the boundary with The Grange.

The Molesworths confess to being great scavengers, many unconsidered trifles have found positions of utility and beauty. The current organic vegetable garden is notably productive and enjoyable, the old concrete base of a small greenhouse is used for tables and chairs. The current greenhouse is an assemblage and the polytunnels were rescued. The exquisitely woven hazel arbour for training beans is a work of art in its own right.

There is an inherent generosity of spirit in the attention to protecting nesting birds and small mammals (I was thrilled to see a dormouse nest in one of the sculptures); the elderly rescue donkey quietly grazes, the laying hens cackle to themselves, and the tortoises make ponderous progress. The topiary includes a bird on its nest atop a hedge. Charlotte's passion for gardening and plants was nurtured by her mother. Some plants date back to her childhood; a delight in yew trees, old fashioned roses, cottage garden plants as well as geraniums, box hedging and, of course, topiary are manifest. Any gardener will tell you, if you want to keep a plant, share it. Charlotte's sheer genius can be enjoyed at Penshurst (pages 20–23) and on Etchingham station – art and craft skills truly unshackled.

Topiary nesting bird, Balmoral Cottage
William Hogarth wrote that beauty lies in the curve. The curvaceous clipping is so natural that it is almost possible to suspend belief to the point that there is indeed a large nesting bird on the hedge.

Topiary fest, peacock, Balmoral Cottage
A spring haze of clipped-yew peacock, estrada and spiral catches the eye while the trees are waiting to come into full leaf and blossom. All against the backdrop of 'Cherry' Ingram's mature trees.

Table and chairs embosomed, Balmoral Cottage
Profusion is one way to celebrate the best of English gardens. This is more than a tapestry of textures, it invites you to come and just sit.

COLWOOD HOUSE, WARNINGLID

the High Weald's 'Come into the Garden, Maud'

The Building News, 31 October 1884 published a watercolour of 'Colwood, near Cuckfield, Sussex, the seat of Lord Justice Bowen. Halsey Ricardo. Architect.' A half-timbered Victorian pile on local stone with dominant Tudor-style brick chimneys. This early work by Halsey Ricardo was in fact an extension in 'Norman Shaw Tudor' style to a farmhouse encased in a red brick Georgian manor for Sir Charles Bowen, Lord Justice of Appeal from 1882 to 1893, created Baron Bowen of Colwood in 1893. In 1890 the Rt Hon. W.E. Gladstone and his wife paid a weekend visit to Slaugham. Warninglid is in the same parish, crowds had lined the route from Haywards Heath station and Colwood had bunting festooning its gates. They were marking their 51st wedding anniversary, accompanied by the Bishop of Peterborough, Sir Alfred Lyall and Lord Justice Bowen. As he left, Gladstone said: '*My visit being of a private nature I must forbear to tread upon the toes of any political opponents. This is, indeed, a beautiful part of the country and I have much enjoyed my visit.*' Colwood became a school before being divided into 11 parts in the 1980s. The turning circle has the fourth largest Himalayan cedar (*Cedrus deodara*) in West Sussex.

In 1987 Rosy Brenan moved into the main block of Colwood House with 12.5 acres of the original 20. The house and gardens were neglected, and by October the Great Storm had uprooted over 40 trees, 20 of which fell over Cuckfield Lane, a further four over the drive and two on the house. One fallen redwood is sending up new shoots from its base. Despite the trees being dwarfed by the underlying sandstone four are Champion and/or the largest in Sussex. The back of the house is south-facing, supporting a magnificent fig, with a deep terrace from which to admire the gardens. Throughout, architectural salvage found by Rosy has been reinvented to great effect.

The house is divided along its west side from whence the almost imperceptible boundary runs into the landscape, widening as it descends. To the east is a seemingly traditional herb garden with a central font, formerly for spa water in Tunbridge Wells. The third largest *Liquidambar* in West Sussex marks the start of the parterre planted with hardy geraniums and ageratum; it is in symmetry with the swimming pool beyond (the windows in the pavilions came from Brighton's demolished Regent cinema). A dense border of hydrangeas leads to the Rose Walk which is structured with blacksmith-made frames for rambler roses that make a spectacular display in June and are underplanted with lavender, leading to the Beech Lawn. A long border of tall plants hides the tennis court which lies in front of the Rose Garden planted with David Austin English Roses. A statue of a satyr entwined with a woman is engulfed by the rose 'Rambling Rector' under the watchful gaze of the third largest Gingko in West Sussex. Leading to the Temple of Venus are two rose granite pillars which were once part of a Victorian chemist shop front in Brighton. A shady walk leads past almonds, quinces and cherries that provide blossom from late February onwards, into the front gardens which have Ghent azalea plantings.

Below the parterre on the east side, a small series of gardens lead down to a little glade of the four seasons, and a fruit garden with olives, persimmon and a pineapple guava. This area is attributed to William Selkirk who died at the age of 91 in 1961 after falling off the billiard table on New Year's Eve – so unbelievable it must be true. There is a cut-turf labyrinth with a stone griffin at its centre. A grass path runs alongside a wisteria-covered pergola towards the lake where an ironwork bridge leads to the island and Temple of Venus – a good vantage point from which to look back at the house and across the gardens. Less intensively planted, there is a fine collection of rhododendrons, a giant chess board and a small orchard of apple and pear trees – perambulating series of walks with interesting and curious incidents. Rosy has two gardeners who each work two days a week keeping the trees, shrubs, flowers and lawns in excellent shape.

Naturalised flowers, Colwood House
Colwood House can be seen at the top of the gardens, their formality dissolving into naturalised flowers.

Rose garden and pavilion, Colwood House
The white 'gazebo' that sees all was originally an aluminium fish & chip stall in Cardiff. Far more upmarket than it sounds, it arrived in kit form and was rebuilt, the wooden seat backs coming from Peckham station.

(Previous pages) Lake and Temple of Venus, Colwood House
The Temple of Venus invites you to sit and contemplate the gardens and house. The lake has been newly restored.

Terrace to cedars, urns and croquet lawn, Colwood House

In Victorian style an immaculate croquet lawn runs by the terrace. The view beyond and down to the lake is punctuated by mature trees such as a Himalayan cedar (*Cedrus deodara*) – the third biggest in West Sussex.

MERRIMENTS GARDEN

ideas to take home

What inspires someone to garden? Plants or growing displays? At Merriments you arrive through the excellent nursery, passing all the garden tools that you need and a lot more that you will desire. A gate leads you into the four acres designed to inform aspiration and skill. In an echo of Robinson's daffodils at Gravetye, the annual mass planting of tulip bulbs is growing year by year, starting in 2018 with 8,000, by 2019 it was 21,000. The cupped flowers in an array of shapes and shades create a colourful swell across the beds, the garden was originally designed with colour-themed beds of mixed planting which are gradually being reinstated. Following the tulips come Indian hyacinths, the alternative descriptive name for *Camassia*, ideal for a damp site, they have naturalised into tall blue swathes. In 2017 Lucy Cross bought the gardens from the Weeks family; Mandy Weeks had designed the original layout with her husband Mark Buchelle for their family business including landscaping and a nursery. Merriments has built up a reputation as specialist growers of perennials, the late season herbaceous perennials reach their peak creating a colourful spectacle in late summer and autumn.

As part of the gradual redesigning of the gardens, French drains were installed during the winter of 2019/2020; water is being directed underground into the Monet pond complete with eponymous bridge. In turn this feeds the Willow Pond via a pumped stream; it is the original water system the garden was designed around. The Asian Garden was built with pond and waterfall in 2018, its bridge is clear walled so that there is no visual barrier between the visitor and the naturalistic rocks that form the banks. Water is channelled in a series of rock edged steps, each providing a small pool between the falls, framed by swamp cypress. In 2020 the former formal box-edged beds fell prey to box blight so are in the process of being redeveloped. The Monet theme is extended by a series of Giverny-inspired arches leading to the Exotic Garden. The Fern Garden is far from formal, it has an extraordinary range of shade-loving plants colonising under a canopy of tree ferns (*Dicksonia*).

Merriments enjoys the status of being an RHS Partner Garden – it has received funds from the LEADER Programme to help update its retail outlet and the restaurant. Wherever possible the plants are raised on site or by British nurseries – there is an excellent selection of ornamental trees. Look out for the pink pots as they are recyclable. Four acres is undoubtedly larger than most private gardens, however, the planting ideas can truly be decanted into a home plot.

Spring Garden, Merriments
The tremendous range of hellebore varieties is clearly illustrated in the dappled spring shade at Merriments. The flowers of snakeshead fritillaries and grape hyacinths also weave additional colour amongst the plantings. (Courtesy Merriments Garden)

New Asian Garden, Merriments
Water takes centre stage in the Asian Garden, with the stepped waterfall feeding the pool that in turn reflects the Japanese acers. In all, it exudes a sense of tranquility. (Courtesy Merriments Garden)

Autumn perennials, Merriments

The bright pink flowers of *Schizostylis* will brighten up a flowerbed or border from autumn into the winter. The tall purple heads of the *Verbena bonariensis* add architectural notes and the bright yellow *Rudbeckia* add further bright notes amongst the autumn perennials. (Courtesy Merriments Garden)

LEONARDSLEE

where colours run deep

In terms of shaking off the historical shackles should Leonardslee be in this chapter? Its creation and plantings are world class, inspirational and global, but were disappearing. The shackles – that they were achieved under very different circumstances – means the 21st century presents a make-or-break situation. The remaking to high standards has been unfettered by cost and resolutely commercial, but without the loss of great crested newts. From 2019 Leonardslee Lakes and Gardens has opened throughout the year, welcoming visitors through the aptly named Crystal Palace, a new glass structure that serves as a plantfest entrance. Ten miles of new stone and hoggin paths have been laid across the 240 acres. Leonardslee's reincarnation is thanks to Penny Streeter who acquired the estate in July 2017. Her son Adam is the general manager and lives on site. The Streeters also managed to buy the mansion which the Loder family had sold in 1984. It is a solid Italianate house in gold-flecked sandstone. The Camellia Walk near the mansion is unusual in that it includes thriving palm trees: there is the blood-red *Camellia* 'Takanini' (bred in New Zealand) and the more traditional *C.* 'Lady McCulloch', the tender former Spider azalea *Rhododendron stenopetalum* 'Linearifolium' which has long narrow leaves with deeply divided pink flowers, and an interesting group of *Pieris* including *P. formosa* 'Wakehurst' with its vivid red young foliage that contrasts beautifully with the glistening white flowers.

Stephen Herrington left Nymans to take up the post of Head Gardener here in October 2019. There are 18 gardeners, four of whom are arborists. A former gardener provided a handwritten list of 1,600 plants, and a further 10,000 have now been catalogued. The initial planting of Christmas trees has been removed from the Deer Park and Pinetum. Membership of the International Conifer Conservation Programme links Leonardslee with Bedgebury and Wakehurst. One programme with the latter is sharing seedlings of the critically endangered Sicilian fir (*Abies nebrodensis*); a native of the Nebrodi and Madonie mountains in Sicily, a mere 21 trees survive. In 2019 the waterfall was restored in the famous Rock Garden, built by James Pulham and Son. As they did in Sheffield Park, local sandstone was mixed with pulhamite. The rocks host a dense colourful patchwork of the evergreen Japanese Kurume azaleas which were originally introduced by 'Chinese' Wilson. The remarkable compact dome-shaped *Rhododendron yakushimanum* has silvery young growths and glossy green leaves with deeply brown undersides. It is only found on the windswept mountain peaks of Japan's Yakushima Island. These are just two of many azaleas and rhododendrons which, with three groups of palm trees (the clump-forming varieties of Chusan palm), are unique to Leonardslee.

The Leonardslee landscape lies on the forest ridge, the steep slopes mirroring the Himalayas. It is essentially frost-free and, like Borde Hill, has a wide range of microclimates. In 1852, with excellent rail links from nearby Horsham, W. Egerton Hubbard bought the estate of 1,000 acres including 30 acres of beautiful parkland. Three years later he named it Leonardslee and built himself the Italianate mansion. The name Loder is key to the development of High Beeches, Leonardslee and Wakehurst Place, as well as its close association with other collectors of rhododendrons, azaleas, camellias, magnolias and acers. In 1878 Sir Edmund Loder married Marion Hubbard. In 1888 they acquired Leonardslee, enabling Edmund to pursue his great interest in rhododendrons and exotic conifers. He crossed *Rhododendron fortunei* with *R. griffithianum*, resulting in the hybrid *Rhododendron* 'Loderi' in 1901. In turn this produced many large-flowered and fragrant varieties; arguably his finest is 'King George', but others include 'Pink Coral', 'Sir Joseph Hooker', 'Venus' and 'White Diamond'. Two *R.* 'Loderi' varieties were crossed ('Sir Edmund' with 'Snow Queen') to create *R.* 'Mrs. C. Whitner', named for the wife of one of the gardeners and which was presented with an Award of Garden Merit by the RHS in 1963. Predating Sir Edmund and

***Cornus kousa* fruits, Leonardslee**

Cornus kousa fruits which Stephen Herrington hopes will be served in the restaurant. They should be bright red and soft to the touch. The flavour is said to be reminiscent of mango and pineapple, the skin is edible but tough, and you need to spit out the pips.

Ponds and rhododendrons, Leonardslee
A landscape of textures, colours and reflections looking down across the central series of ponds. In the foreground, the naturalised rhododendrons in an exotic arboreal setting. The bright yellow flowers of *Rhododendron luteum* wafting their scent across the paths. (Courtesy Leonardslee Garden)

still surviving is a *R.* 'Jacksonii', one of the earliest raised hybrids dating to 1835. Tolerant of industrial pollution and with an iron will to thrive, as a slow-growing domed-shaped rhododendron it has a lot to recommend it for domestic gardens with acid soil. The glorious *R.* 'Loder's White' was raised in 1884 by the noted rhododendron specialist Harry Mangles of Littleworth Cross, Surrey. Five years later he invited Gertrude Jekyll over to see another new rhododendron, and who should also be there? His young architect Edwin Lutyens, the start of Lutyens and Jekyll's fruitful friendship.

Sir Edmund ran the estate until his death in 1920. His son Robert was killed in World War I, leaving the infant Giles whose mother Muriel then maintained the estate. In 1945 Giles Loder took over with his wife Marie née Symons-Jeune, from a family of noted landscape designers. In 1952 a Coronation Garden was created in honour of Queen Elizabeth. Young examples of all the Loderi offspring were planted, including 'King George' and the now extremely rare *R.* 'Spearmint'. Amongst the plantings are the gloriously scented yellow-flowered *R. luteum*, and the enormous-leafed as well as fragrant *Magnolia macrophylla*. *Rhododendron* 'Leonardslee Giles', named in his honour, was raised here in 1948. In 1981 the couple retired, handing over to their younger son Robin. Towards the Mossy Ghyll are two oaks: a Hungarian or Italian oak (*Quercus frainetto*), a favourite of Sir Robin Loder, with a Macedonian oak (*Q. trojana*).

As at Gravetye, the Head Lake originally served ironworks. Crossed by the Clapper Bridge, it feeds a north–south belt of ponds that reveal and conceal the landscape – Top Ponds, Mossy Ghyll Pond, Engine Pond, Waterfall Pond and finally New Pond. Some of the oldest plants are in the Upper Dell whose outlook includes the three top ponds. Mature rhododendrons such as the Exbury hybrid 'Golden Dream' with deep-cream flowers flushed and shaded pink fading

(Previous pages) Pinotage vineyard, Leonardslee
Several vineyards have been planted around the margins of the High Weald. Planted in 2018 the Leonardslee Pinotage vineyard was the first in the UK. Cinsaut and Pinot Noir grapes were first crossed in the gardens of Professor Abraham Perold in South Africa in 1925. Its thicker skin makes it more resistant to rot and disease. (Courtesy Leonardslee Garden)

Pulhamite rockery, azaleas in flower, Leonardslee
Not a Wealden rock outcrop but Pulhamite stone, made by the company James Pulham and Son. They mixed it with the local sandstone spoil when the croquet lawn was created. The effect is dazzling and dramatic. (Courtesy Leonardslee Garden)

to white, and the large-flowered 'White Glory', are mature specimens. Of both historic and actual stature are a *Magnolia campbellii* potentially the largest and one of Britain's original dawn redwoods (*Metasequoia glyptostroboides*), which was sown here in 1948. There are a further six specimens that survive elsewhere in the gardens. The combination of camellias, rhododendrons, azaleas and acers, to name but a few, makes this one of the most colourful landscapes in Britain with reflections in the water doubling the effect. There is even a tricolour European beech (*Fagus sylvatica* 'Purpurea Tricolor'), a smaller form that is resistant to deer; the edges of the leaves are a clear rosy pink with white variegations on a copper beech background.

At the time of writing, a new area, to be dedicated to the flora of Australia, is working its way off the drawing board. The first setting would have conflicted with the surrounding 300-plus oaks whereas the new development at the entrance allows greater flexibility. That said, there are already many established Australasian trees and shrubs. The design is by writer and landscape architect John Wright who lived in Australia and worked in Tasmania. He has selected hardy gums such as *Eucalyptus debeuzevillei* and *E. niphophila*, and the splendidly named Wally's wattle (*Acacia pataczekii*).

Albino wallaby, Leonardslee
As well as the extraordinary plant introductions, Sir Edmund Loder introduced a wide range of animal species, notably wallabies in 1889. They are free to roam but have a safe enclosure for breeding complete with a Pulhamite shelter. (Courtesy Leonardslee Garden)

CHAPTER FIVE

Sheer folly – landscapes with a ripping yarn

Whether you have read Frances Hodgson Burnett's *The Secret Garden*, listened to a serialisation, or watched any one of the many adaptations, it is a splendid evocation of the restorative powers of an English garden. Although the story is set in chilly Yorkshire, the inspiration came when the author discovered the walled garden at Great Maytham Hall, where she lived from 1898 until 1907. In 1909 Edwin Lutyens was commissioned to rebuild the house and lay out new gardens for H.J. Tennant. This included opening up the walled garden whose secret door no longer leads to anywhere. Although now divided into flats, the house and gardens remain fine exemplars of Lutyens' work. The gardens are regularly open and worth a visit even if the ripping yarn has been unravelled.

Erasmus wrote that '*folly is the seasoning of pleasure*' – the village and surroundings of Brightling's landscapes are peppered with quite a selection of folly. Sheep graze around the Pyramid in the churchyard of St Thomas à Becket where the mortal remains of 'Mad' Jack Fuller, squire and politician, were laid to rest in 1834.

The 6,500 acres of Ashdown Forest stretch across Surrey, Sussex and Kent. It is a Site of Special Scientific Interest in an Area of Outstanding Natural Beauty, and a site of literary pilgrimage. Pooh Corner is south of Hartfield, the Pooh sticks bridge is a mile from the Pooh car park near Chuck Hatch on the outskirts of the Forest. There are no fears of heffalumps when you enter the Hundred Acre Wood to seek out the Enchanted Place or Gill's Leap; you will then find Eeyore's Sad and Gloomy Place, Roo's Sandy Pit and the North Pole.

The ripping yarns that follow start with Henry VIII and his pursuit of Anne Boleyn at Hever Castle, in contrast with Hammerwood Park whose elegant house is reputedly the key folly. The great storyteller Rudyard Kipling found peace and solace at Bateman's while an early owner of Twyford House was an uncle of Gothic horror master Matthew Lewis. Lastly, two houses united by art and nature: the picturesque folly of ruined Scotney Castle overlooked by the Elizabethan Revival house built at the dawning of the Victorian era.

Willow stump sculpture of Kipling, Bateman's
Steve Andrews carved this enigmatic sculpture of Rudyard Kipling reading a story out of an old willow stump in 2017. This is doubly appropriate as it overlooks the lily pond which was built using the monies received from the Nobel Prize for Literature.

MOATS

a symbol of status

The fear of attack continued after the Norman Conquest. Initially castles, wooden or stone, were encompassed by rectangular, round or square moats, making them easier to defend. As peace reigned, their ornamental value as mirrors around grand houses ensured their continued creation. Manorial households would move between their properties to benefit from the harvest of each thereby not depleting their resources. Well-managed moats were a constantly renewable resource, at intervals emptied and excavated, the spoil spread on the land ensuring increased fertility, minimising the health hazards for the household and the stink. Moats were populated with pike and carp which scavenged the refuse, added elegant ornamentation, and provided food for the table. The remnants of the moat at Great Dixter are in two parts: the Lower Moat yielded up the old cellar door which provided the template for the Lutyens restoration and extensions. It is now a pleasing water feature alongside the nursery.

Moated old Scotney Castle
The most exciting story emerged from an icy Scotney Castle moat: from 1591 a Catholic priest, Father Richard Blount, had been concealed in a hole by the staircase. He was being hunted by the local Justice of the Peace. In 1598 he was detected, fled and plunged into the icy moat, and swam to his servant and a horse to make their getaway. (zawisak/Shutterstock.com)

Moat, Tudor Village, Hever Castle
Across the water lies the Anne Boleyn's Orchard which has ornamental apples and cherries (*Malus* and *Prunus*). Their blossom makes a fine spring display underplanted with bulbs.

(Previous pages) Lichen-encrusted loungers, Twyford
Two lichen-encrusted loungers at Twyford, proof of how clean the air is on the High Weald. Once the seating has become an art form, those seeking relaxation must look elsewhere.

Heath, Chelwood Vachery

A view from the heath near Chelwood Vachery looking towards the setting for A.A. Milne's Hundred Acre Wood. The splendidly named Chelwood Vachery could be described as a naturalised forest garden – a popular walking area.

HEVER CASTLE

romance and riches

The loves and wives of Henry VIII are amongst England's great stories, especially Anne Boleyn – sexy, intelligent and mother to the great Elizabeth I. Unabashed, Henry later installed his fourth wife, Anne of Cleves, here for 17 years after their divorce in 1540. No wonder William Waldorf Astor was drawn to buy the Hever estate in 1903. On a site dating back to *c.*1270, the 'castle' or moated manor house was built in 1384 by Sir John de Cobham. Naturally, it was not large enough for his guests, so a new Tudor 'village' was built in sandstone quarried from the same seams at Ide Hill. He then turned his attentions to the 30-acre gardens and the creation of a 35-acre lake. At every turn there was the opportunity to display his superlative collections.

William Waldorf Astor's great grandfather encapsulated the American Dream: Johan Jakob emigrated from the German town of Walldorf, near Heidelberg, to America in 1783. He started to trade fur in the north-east, buying from Indian trappers. Within 20 years he owned a fleet of a dozen merchant vessels. He exported his furs to Europe and the Far East and imported manufactured goods and tea. He invested his profits in New York real estate, and in 1848 he died the richest man in America. His descendants branched out into politics, hotels, magazines, newspapers, racehorses and agriculture. From 1882 to 1885 William Waldorf was American Minister (or Ambassador) to Italy, where he acquired a connoisseurship of classical antiquities. In 1890, with a fortune estimated at $100,000,000, he left America and became a naturalised British citizen. By 1893 he had bought Cliveden in Buckinghamshire from the Duke of Westminster. In 1901 he fell in love with the romantically ruinous Hever Castle; two years later he bought it with 640 acres. Not only did Astor then lavish a fortune on Hever, he was also a generous benefactor. In 1916 he was created the first Baron of Hever Castle, a year later Viscount Astor.

The first major landscaping project was to divert the course of the River Eden and the public road to accommodate the 100-room 'Tudor Village' designed by Frank L. Pearson. Close to the castle Pearson designed Tudor gardens: a yew maze which remains a hugely popular exercise, a splendid set of topiary chessmen overlooking an astrolabe and the castle across the moat, and between them are Anne Boleyn's knot and fountain gardens. The approach from the east to the castle is now planted naturalistically as Anne Boleyn's Orchard, more dramatically to the west is the great chestnut avenue which traverses Sixteen Acre Island between the River Eden and the lake. Anne Boleyn's Walk was created to run from the top of the Golden Stairs, parallel to Tainters Hill Field and Pokeridge Meadow, to Rider's Croft by Splash Water on the Lake. Introduced more recently, and requiring a certain agility, is the water maze, designed to soak if you misplace your steps.

The formal gardens to the east of the castle lead to Pearson's flamboyant Italian Garden. The Pompeiian Wall lining one side was designed to display Astor's remarkable collection of classical sculpture and collections of architectural salvage amongst lush planting; balanced by a massive pergola opposite, swathed in wisteria which shades magnificent pots of camellias. All of this leads to a loggia and lake that could be a setting for Alma-Tadema. In 1903 Walt Disney was only two years old when Astor acquired Hever, but such is its history and drama that it has become event central. The Italian Garden is frequently used for concerts and spectacles, which detracts from its former Roman ruins incarnation.

In and around the gardens, children can climb Tudor Towers, explore Acorn Dell, or quite simply burn off excess energy in the playground. Visitors can picnic, snack or dine as well as walk – away from the crowds – and look around inside the castle to admire the gardens from the Tudor windows and contemplate the past.

Italian Garden, side garden, Hever Castle
One of the side displays in the Italian Garden, columns, a well-head and terracotta pot set on the wall beyond. The purple hues of the red valerian (*Centranthus rubra*) in the foreground are repeated on the wall and in the *Acer palmatum* 'Dissectum'.

Topiary walk to Hever Castle
The topiary walk which leads to the castle echoes the historical association of Anne Boleyn and her daughter Queen Elizabeth I. Today's wide approach shows the huge popular appeal of the castle and its gardens.

Rock garden, Hever Castle
Behind the Italian Garden are a series of rock gardens created in Pulhamite artificial and locally occurring stone.

HAMMERWOOD PARK

enigmatically cryptic

Following on from Erasmus' definition that folly is the seasoning of pleasure, David Pinnegar, the owner and saviour of Hammerwood Park, will point out to you that the gods look after children and drunken men. In an age of folly-building he argues it was the house itself that offered a journey to the dawn of time, entry was through a side door. Benjamin Latrobe prepared a perspective drawing of his proposed design for Hammerwood *c.*1791/92, commissioned by John Sperling, his portrait shows him standing in the portico in a to-the-manner-born stance with his hunting dog and gun. The size of the columns on the Greek Doric end pavilions are reminiscent of those of the double temples at Paestum, formerly Magna Graecia, modern day southern Italy. In Gainsborough style, his wife Harriet is sitting in the shade of a tree with their children in a celebration of English nature. In his diary Latrobe's brother, Christian Ignatius, wrote: '*Nature has done a good deal for Mr Sperling with low and high woods, vales, runs of water, springs, etc but a little assistance from Art is wanting to render this as delicious a spot as any in the kingdom.*' Meanwhile Henry Fox-Strangways, Lord Ilchester, wrote to Sperling: '*So nice to hear that you, the wife and family are now enjoying the pleasures you came there to seek.*'

By 1791 the American, French and earlier British agrarian revolutions had taken place, all during the reign of George III (styled the 'farmer king'), and the Industrial Revolution was in its infancy. In the letter quoted above Lord Ilchester continued to write in detail about different types of plough, chaff-cutter, and as to whether to employ an overshot or undershot water wheel. One need was to pump water up the hill for a water closet.

In 1865 the second floor above the pavilions, and the third over the house centre, were designed by the Greek Revival architect S.S. Teulon, described as the most exclusively aristocratic of mid-Victorian architects who rejoiced in complex plans, so a fitting choice. He was commissioned by the new owner, successful banker Oswald Smith. The original inscription in Greek: 'Of John Sperling's mansion. The architect BH Latrobe built the first portico in 1792AD and the 2nd year of the 642nd Olympiad' remains valid because the addition is virtually seamless.

The 60 years between 1921 and 1982 brought desolation, culminating in ownership by Led Zeppelin. Into the fray stepped a 21-year-old David Pinnegar, who has spent his inheritance resurrecting Hammerwood to become once more a site of folly and intellectual gymnastics. The landscape falls to the sinuous lake at the bottom of the hill before climbing away across the High Weald. Latrobe claimed to be close to Humphry Repton, whose gently picturesque style is evident. The former Victorian gardens are recovering from their draconian haircut, creating a formal juxtaposition and light woodland by the house while mature trees shade the drive. A canvas for film locations and photoshoots – a story beyond fiction.

Drive and trees, Hammerwood Park
The long tree-lined approach to Hammerwood passes a cricket green until swinging into the drive and meeting the parkland setting of the house.

(Following pages) Portico pillar to landscape, Hammerwood Park
Not just a glorious parkland view over the High Weald. The stout column of the portico is based on those of the Temple of Apollo on Delos. The landscape at Hammerwood was at the forefront of the agricultural revolution and masonic ritual. The masonry of the porticos and the house could be described as poetry in stone.

Park facade view, Hammerwood Park
The park incorporates Repton's picturesque principles of mirroring a landscape painting, with the house and surrounds as foreground, and sinuous landscape with trees as middle ground playing to a dramatic background of forest.

TWYFORD HOUSE

crescendo and arpeggio

Rather than a garden of rooms, this is an operatic experience, scene opening into scene, with the occasional Wealden chorus from the landscape beyond. Jenny and Clive Whitmore love opera (she suggested La Calisto being acted out here – nothing like Jupiter, Mercury, Diana and Calisto to play out love and transformation). Jenny narrates the story introducing the characters who have designed, supplied or crafted the elements that make up their garden. The history of the house dates to William Luther Sewell (1749–1832) whose grave monument can be seen in St Margaret's church, West Hoathly. He was executor and trustee for his nephew, Matthew Gregory 'Monk' Lewis, whose *The Monk* was a hugely popular Gothic horror novel. The Ashdown Forest Dispute of 1876–82 raged around Twyford, the area described as a '*dreary waste of beggary and desolation, a few miserable cattle starve on it.*' Interestingly, it resulted in the continuation of common rights. There is a delightful story of children from Twyford picking flowers for cricket matches at Sheffield Park.

The house was extended in 1917, its division in 1976 created two domains in search of a garden; in 1997 the Whitmores arrived. They called in the garden designer Harriet Scott. Her first advice was to look at their half as a whole and create gardens accordingly. She was so successful that its former status is never apparent. At the front, *Photinia* planted in generous boxes disguise the division. Harriet achieved borrowing the landscape on a dramatic scale, enhanced by her advice that flowers should be avoided towards the margins of the garden as they would compete with the glorious views. Le Vieux Logis in Tremolat, Dordogne is a culinary and horticultural joy with box balls rolling out across its gardens – Jenny adopted the design as a pleasing transition from her lawns to the landscape. The local Wych Cross nursery had the perfect 'pebble' shapes in stock. Jenny feels they echo the hills and rounded treetops. To this formality were added fastigiate trees and blocks of beech. The play of light and shade constantly changes the scene. Julian Furness of The Garden Trellis Company designed trelliswork that frames the garden from the house along the party boundary – simple architecture at its best. Stone steps take the visitor almost to level ground, where a simple parterre of clipped box infilled with shaped santolina acts as a decorative finale.

A long drive leads towards the house, and the white trunks of birches draw the eye to an enclosed garden with mirror pool enlivened by two red juggling figures made of iron, created by Alain Prillard. Prillard is based in Cajac, his motto is '*Faisons un rêve*' ('Let's make a dream'). The French connection continues with the garden areas having French names such as Jardin mystère. Ahead lies a gate that takes the visitor straight into the garden with a long well-planted border curving towards the house punctuated by openings into further green scenes. A flagged terrace surrounds the house, providing plenty of space to sit or eat. Harriet Scott countered the advice to plant in threes, triple it and gain impact and exuberance. The long borders are enclosed by close-clipped yew hedges with an arrow-straight gravel path punctuated by clipped-yew cones. Hornbeam (*Carpinus*) pillars from Tendercare Nurseries contrast with a 40th wedding anniversary present of voluptuous *Hydrangea* 'Annabelle'. *Hydrangea* 'Limelight' is used to line the partition hedge, and is underplanted with masses of bulbs. Path profiles vary, curving grass from the summerhouse where the roses 'Rambling Rector', 'Félicité-Perpétue' and 'Paul's Himalayan Musk' embower the vista.

An unused ha-ha was an ongoing problem. Eventually, Jenny raised the level, complementing the stones with clipped large-leafed box, and disguised the fence with an allée of high, clipped *Photinia*; the straight path leading to the Wendy House is pea gravel, in winter Jenny covers it with a fine mesh which prevents discoloration. Beauty meets utility in the no less than four handsome compost bins whose contents help to feed the soil, and a construction masterstroke, the steps in the brick paths have a central flat brick for wheelbarrows.

Topiary pebbles, Twyford House
The topiary box pebbles roll down the lawns to call in the country, while fastigiate hornbeams guard the gateway, their manmade shapes contrasting with the oaks, chestnuts and conifers in the landscape. The iron fence closes the garden but leaves open the view. It was made by the Gravetye blacksmith. (© FMcLeod)

Statue, Twyford House
The temple is positioned due west so it becomes bathed in late afternoon sunshine. Its Italian formality, swathed in roses, contrasts with the oak tree trunk, shaded pastoral statue and the High Weald beyond. (© FMcLeod)

Autumn tints and temple, Twyford House

Amid autumn colours on a dull early morning, the oak dominates. The beech hedge adds a deeper russet, tucked in by the moss-clad wall a domed lavender adds further texture. The High Weald lies in the mists beyond.

SCOTNEY CASTLE

Bless'd too is he, who, 'midst his tufted trees,
Some ruin'd castle's lofty towers sees;
. . . Its roof with weeds and mosses cover'd o'er,
And honeysuckles climbing round the door;
White mantling vines along its walls are spread,
And clust'ring ivy decks the chimney's head.
The Landscape: A Didactic Poem by Richard Payne Knight

Scotney Castle's storybook setting would have thrilled and delighted Richard Payne Knight. *The Landscape* was published in 1794 exciting much debate on the merits of the picturesque and aesthetics. Architectural romance and floral fecundity meet here with romance, ruins and rhododendrons, around a moated castle, across an excavated quarry, and overlooked by a patrician house.

Hussey family money had been made in the iron industry in Worcestershire. They moved to Sussex in the early 1700s. In 1778 Edward Hussey purchased Scotney Castle and its 1590 south wing, Ashburnham Tower and the estate. Tragically, he committed suicide in 1816, and his son Edward II survived him by one year; his wife Anne chose not to use it. In 1835 Edward III decided to build a new house. He commissioned the young architect Anthony Salvin, and in the course of 33 meetings they finetuned the house's aspects with the main reception rooms placed along the south and east fronts to benefit from the painterly views across the gardens to the ruined castle. The old quarry was excavated for its streaky golden sandstone and building work started in February 1837. The fine Elizabethan Revival house was completed in 1843. It sits on the highest ground with a ha-ha to the west so that there are open views to the pastures. Accommodation was left in the castle and old house, the south wing was occupied by the head gardener, George Wells, and then the Estate Bailiff until 1905. The ancient trackway to the castle and its 15th-century house survives and provides a rural approach.

Edward Hussey approached artist and landscape gardener William Sawrey Gilpin (nephew of the picturesque supremo the Reverend Gilpin) to design the landscape around the Bastion. The stone balustrade acts like an opera box from which to admire the scenery culminating in a picturesque folly, the now ruined Castle. Once the building was complete Edward III, and after his death his son Edward IV, were content to enjoy the life it offered, the latter adding plants. During World War II The King's School, Rochester were evacuated here. Edward Hussey's nephew Christopher inherited in 1952. He had worked at *Country Life* for over 30 years, was at the forefront in saving historic country houses from falling into ruin, and wrote a biography of Edwin Lutyens. He had proposed to Elizabeth Kerr-Smiley in the grounds, and they married in 1936. They dedicated themselves to sensible modernisation and sensitive attention to the landscape. The castle herb garden was designed by Lanning Roper around a carved Venetian well-head. On Christopher's death in 1970 he left Scotney to the National Trust, at which point the gardens opened; on Elizabeth's death in 2006 the house was opened. A tour of the interior is essential for the views across the landscape and the eclectic portraits and cartoons. In the library, Hogarth shares the shelf with *Kentish Grasses*, *Ornamental Gables*, *Greenwood's Map of Kent*, *The Worthies of Sussex* and John Evelyn's *Sylva, or a Discourse of Forest-Trees.*

The walk from the car park passes by one of the best tree swings in the National Trust, with open views across the High Weald. A former templar entrance, now bricked up, to a substantial walled garden hoves into sight.

Scotney Castle long view
The Darell family lived at Scotney Castle for 350 years. In 1580 they built a south wing onto the Ashburnham Tower. Fifty years later William Darell demolished much of the castle and recycled the materials to create a new three-storey east range whose walls now support roses. (Hannah Denski/Shutterstock.com)

Victorian house, Scotney Castle
The south terrace looking towards Salvin's house, where the fountain in the centre of a quatrefoil pool gently plashes. The splendidly realistic cat is actually in stone. (vanhurck/Shutterstock.com)

Outside is an informative board on what to look out for. Make sure you go in as it is filled with good things. Onwards past the stables – now café and shop – to enter the front courtyard and the façade of the battlemented house inspired by Salvin's Northumbrian roots. The Great Storm of 1987 shattered the picture that had been so carefully nurtured. Cedars and cypresses fell, but in the long term the scoured canvas provided new opportunities. At the turn of the 21st century there was an overwhelming call to reinstate the 1837 picturesque view created by Edward Hussey III. The plants in the Quarry and Main Lawns obscured the picturesque views to the moat and castle, and the quarry rocks were lost. A detailed survey was undertaken to record significant surviving plants. In 2018 funding was secured to clear the overgrown, thuggish *Rhododendron ponticum*, whilst encouraging one of Christopher Hussey's pride and joys for which Scotney was famed – *Kalmias*. He wrote in *Country Life*: '*In June* Kalmia latifolia *with its calico pink flowers is a rare and noble sight; there are few gardens where such large specimens*

Autumn, Scotney Castle
The view across the moat to the romance of old Scotney Castle. Framed by the leaves of colourful introductions that lead the eye to native High Weald trees beyond. (Patrick Wang/Shutterstock.com)

flower so abundantly.' New *Kalmias* (aka mountain laurel or spoonwood – Native Americans used them to make spoons) are being planted, plus lower-growing rhododendrons which will not compromise the view. The grass path is being reinstated. The journey through the Quarry had been compromised by an overgrown carpet of the American wintergreen (*Gaultheria*), which was removed to expose that key sublime picturesque feature – rugged rocks. Heathers were popular in the early 20th century for giving winter colour, so it was decided to replant along with dwarf conifers based on planting records and old nursery invoices.

From a naturalistic start of snowdrops, primroses and daffodils, April brings the white starry flowers of *Magnolia stellata* and the heavy fragrance of Ghent azaleas; summer displays fuchsias and gentians; as they fade the scene is washed with the blue of hydrangeas. Then autumn, with the fruitfulness of the kitchen garden contrasting with the landscape's dark hues of red and yellow across the foliage of liquidambars, nyssa and Japanese maples.

Mill pond, Bateman's
Park Mill and its ponds were part of the extra land Kipling bought when he acquired Bateman's in 1902. Productive as well as being an adventurous place for his children. An orchard with nine fruit trees has been planted in memory of nine young men who died in World War I. (SuxxesPhoto/Shutterstock.com)

BATEMAN'S

. . . we reached her down an enlarged rabbit-hole of a lane. At the very first sight the Committee of Ways and Means said 'That's her! The only She! Make an honest woman of her – quick! We entered and felt her Spirit – her Feng Shui – to be good. We went through every room and found no shadow of ancient regrets, stifled miseries, nor any menace . . .

The 'Committee of Ways and Means' was Rudyard Kipling's wife Carrie.

It was love at first sight in August 1900 when they saw Bateman's which they bought in 1902. Kipling took on the mantle of English squire, immersing himself in British folklore as avidly as he had in India. Kipling's daughter Elsie Bambridge remembered that he compared the setting of Bateman's to a matching cup and saucer, the valley of the river Dudwell was such a good saucer that its rising waters sometimes lapped the 'Quarterdeck' terraces. Kipling acquired extra land so that his five-hectare (12 acres) garden had privacy within the 121 hectares (300 acres) of farmland, including Park Mill and its ponds. His ripping yarns of India and the British Empire had brought fame and fortune. He noted that Bateman's was *'alive with ghosts and shadows'*, the local sandstone streaked with iron, the red-brick chimneys and mellowed stone mullion windows make it the quintessential English house in the shadow of the Wealden hill.

Like Gravetye Manor, Bateman's was built by a Wealden ironmaster. The date carved into the stone above the porch is 1634, whereas the wrought iron bell pull was a gift from Kipling's aunt, Georgiana Burne-Jones. The original garden walls still gently enclose, but on arrival Kipling found the front walls too low so he planted yew hedges to ensure his privacy. The approach to the front door is in true arts and crafts style, quiet green grass crossed by a straight path made of 'Sussex Marble'. Like The Priest House at West Hoathly, Bateman's has an ironstone slab at the threshold to keep witches out. The interiors continue to be filled with flowers from the garden, and a selection is picked for show in the excellent shop.

So many threads of Englishness are well met in Kipling and his former home and gardens. The gung-ho 'The Glory of the Garden' (published in 1911) used the metaphor of the garden and its glories to galvanise a nation into preparation for military action – the backdrop to the happy childhoods of his son and daughter, as were the slopes of Pook's Hill where the ear is caught by bird song and rapid streams. The garden remained glorious but how changed were his views on the glory of war when less than 10 years later he was appointed Literary Adviser to the Imperial War Graves Commission (now the Commonwealth War Graves Commission). He chose five words from the Book of Ecclesiasticus for Lutyens' Stones of Remembrance 'Their Name Liveth for Evermore' and penned the simple words 'Known only to God' for the graves of unknown soldiers. Both are engraved on monuments in over 23,000 cemeteries, a timeless legacy created in a study full of memories at Bateman's. He had written the words 'lest we forget' in celebration of Queen Victoria's Diamond Jubilee in 1897. Family bliss was shattered in 1915 when their son John was reported missing after the Battle of Loos. His body was not recovered, he was assumed dead. 'My Boy Jack' was not written for him but rather for his part in the horrific loss of so many.

Two of Kipling's original designs remain eye-catching: on arrival you are drawn to the Pear Allée which he designed as a long view towards the house. None of its original spirit was lost in the 2007 restoration, the frame serves its purpose supporting the trained branches as they progress from white blossom to handsome fruits and autumnal leaf colour. The surrounding orchard contains the apple 'Beauty of Bath' which was raised from one of Kipling's original trees, with other apple varieties including crabs, as well as pears, plums, quince and medlar. A large productive vegetable garden and the Herb Border dating from the 1970s supply the tearoom. The Pear Allée gently leads you to a walled enclosure whose central brick path features a former millstone – one of many that adorn the gardens, some are solid millstone grits from Yorkshire, others French burrs held together with local iron bands.

Beyond the house, Kipling designed the rectangular lily pond and rose gardens (his original plans are displayed in the house), which were financed using the £7,770 from his Nobel Prize for Literature in 1907. Alongside the sundial at the end of the rose garden lies Kipling's original gravestone engraved with R.K., which had been in Westminster Abbey until he received a larger tombstone in Poets' Corner. In 2017 Rudyard Kipling reappeared at Bateman's as a sculpture carved out of the old willow tree stump next to the lily pond. In a form that could have stepped out of *The Jungle Book*, the late Steve Andrews of Tribalanimal, a chainsaw carver from Canterbury, portrayed him reading one of his ripping yarns.

Bateman's nestling below Pook's Hill
Rudyard Kipling compared Bateman's house and setting to a cup and saucer. Walking towards the house still encapsulates the perfect English scene and the house appears to grow out of its grounds.

CHAPTER SIX

The plant hunters

Where the wildness pleases takes on a whole new meaning when applied to the men who routinely braved extreme danger and discomfort when seeking new plants. Apart from the arrival of the railways, the High Weald was transformed by the 19th-century drive for innovation and collection. In turn the topography and microclimates of the region provided an adaptable, pleasing, almost untouched canvas. Aligned with innovation and collection, at a time when the sun never set on the British Empire, is the quest for horticultural knowledge. Naturally, there are questionable aspects, however, the surviving botanical discoveries provide living testimony to the many changes in our world's biosphere. This book celebrates English gardens and landscapes created across the High Weald but the history of plant hunters, then and now, is as colourful and blazing as the autumnal hues we enjoy today.

The display on plant hunters at Bedgebury Pinetum starts with **William Dampier** (1651–1715, the first plant hunter to reach Australia), **Jeanne Baret** (1740–1807, one of the very few women), **David Douglas** (see page 167) and their own Plant Collections Manager **Dan Luscombe** (born in 1975). Dan has collected in New Zealand, Australia, China, Vietnam, the United States, Spain, Serbia, Turkey, Japan and Chile for the Forestry Commission. Many expeditions are in collaboration with Kew. The collection of North American forest trees in Horsebridge Wood at Wakehurst includes tulip trees (*Liriodendron tulipifera*) whose seeds were collected by **John Tradescant the Younger** in Virginia in the 1640s. In 1766, in Newfoundland, **Joseph Banks** found and described the Jack pine which was named for him (*Pinus banksiana*). For their great global expedition, Banks and Captain James Cook left Plymouth on 26 August 1768, returning to anchor off Deal on 13 July 1771. Banks and one of Linnaeus' protégés **Daniel Solander** had collected no fewer than 2,000 species, of which 1,400 were unknown.

On 20 December 1771 Carl Linnaeus wrote: '*My pupil Sparrman has just sailed for the Cape of Good Hope, and another of my pupils, Thunberg, is to accompany a Dutchy embassy to Japan; both of them are competent naturalists.*' Wherever 'L.' appears after the name on a plant label it signifies that it was classified by Carl Linnaeus, who commissioned what were known as his 'apostles' to plant-hunt globally. The precocious **Anders Sparrman** (1748–1820) travelled to the Cape of Good Hope to be a tutor to the Resident Governor's children in 1772, botanising in his spare time until he was recruited by James Cook as a naturalist for his second voyage. **Carl Peter Thunberg** (1743–1828), known as the 'father of South African botany' and the 'Japanese Linnaeus', came from the same Swedish region as Sparrman, and became a surgeon in the Dutch East India Company. In August 1775, he arrived at Dejima, a small artificial island occupied by foreigners in the Bay of Nagasaki, a single small bridge connecting them to the city. He was appointed head surgeon and was one of the few to be allowed to conduct limited botanical research ashore. In order to obtain more plants, he traded his knowledge of European medicine with Japanese interpreters for new specimens. In mid-1776 he was finally allowed to accompany the director of the Dutch settlement to the shogun in Edo (Tokyo), collecting as many Japanese plants as he could along the way. In 1784 he published *Flora Japonica*, the first detailed description of Japan's flora and fauna. Many plants to which he gave the epithet 'japonica' were actually Chinese but introduced into Japan, and many plants which he described as living in the wild were actually from gardens. *Thunbergia alata* (Black eyed Susan) and *Berberis thunbergii* are named for him. **Philipp von Siebold** (1796–1866) was posted to Dejima in 1826, where he collected with more freedom than Thunberg and found, amongst others, *Wisteria floribunda*.

Monkey puzzle and rear of Ashdown Park
The mature monkey puzzle tree (*Araucaria araucana*) adds a distinctly Victorian note to the garden front of Ashdown Park. Neither monkey puzzle tree seeds nor acorns from oaks, can be stored in a seed bank as their embryos would dry out, so growing examples of both species is essential to their survival.

Japanese maple in autumn, Goddards Green
Japanese acers offer such a choice in shape and height matched by glorious leaf shapes and colour. It is hard to believe that the first examples were only planted at the turn of the 19th to 20th centuries.

The famous 'exotic nursery' of Messrs James Veitch & Sons of Exeter, and after 1852 the additional Royal Exotic Nursery, Chelsea, rivalled Kew in the field of plant introduction, not least because they had and could make available greater funds. **William Lobb** (1809–1863) was a successful plant hunter in South America and California, introducing the monkey puzzle tree (*Araucaria araucana*) and Wellingtonia (*Sequoiadendron giganteum*) as well as *Berberis darwinii, Desfontainia spinosa, Embothrium coccineum, Nothofagus obliqua, Fitzroya cupressoides* and *Thuja plicata*. In 1861 **John Gould Veitch** (1839–1870) arrived in Japan, then between 1877 and 1880 Veitch despatched **Charles Maries** (1850–1902) who explored Japan and China, introducing amongst many *Acer carpinifolium, A. davidii, Abies mariesii, A. veitchii, Enkianthus campanulatus, Fraxinus mariesii, Hamamelis mollis* and *Viburnum plicatum* 'Mariesii'. **Ernest Wilson** (see below) also collected for Kew. Veitch supplied plants for Borde Hill, High Beeches, Leonardslee and Wakehurst, and to William Robinson for Gravetye Manor.

CHRONOLOGICAL LIST OF 19TH AND 20TH-CENTURY PLANT HUNTERS WHOSE FINDS POPULATE THE HIGH WEALD GARDENS

David Douglas (1798–1834) Scottish botanist who explored the Scottish Highlands, North America and Hawaii, where he died. He introduced amongst many *Pseudotsuga menziesii* (Douglas fir); and *Abies procera* (noble fir); pines *Pinus coulteri, P. ponderosa, P. radiata* as well as *Garrya elliptica,* and *Spiraea douglasii.*

Robert Fortune (1812–1880) Started at the Royal Botanic Garden Edinburgh, and progressed to the Horticultural Society's gardens at Chiswick. He was the first plant hunter to have a Wardian case, to describe tea-growing and chrysanthemum cultivation. He brought back tea plants for the East India Company, and introduced the Chinese gooseberry (renamed kiwi fruit) to New Zealand. When plant-hunting in China (1843–46 and 1848–53) he was able to pass himself off as Chinese. His third expedition in 1860–62 was principally for collecting in Japan. Amongst the many plants he collected: *Akebia quinata, Anemone japonica, Cephalotaxus fortunei, Cryptomeria japonica, Ilex cornuta, Jasminum nudiflorum, Lonicera fragrantissima, Mahonia bealii, Prunus triloba, Rhododendon fortuneii, Weigela florida.*

George Forrest (1873–1932) Between 1904 and 1932 he undertook seven collecting expeditions in Northwest China and Tibet. Introduced more than 300 new species of rhododendron as well as camellias and primulas. Sponsored by Colonel Stephenson Clarke of Borde Hill, from 1912 to 1932 he supplied many plants for Warren Wood, Gore's Wood, The Tolls and the Pinetum: only one *Picea likiangensis* has survived, and the *Rhus verniciflua* was uprooted in the 1987 storm. Other plants include: *Acer forrestii, Abies delavayi forrestii, Berberis replicata, Buddleja fallowiana, Camellia reticulata, C. saluenensis, Cotoneaster lacteus, Deutzia monbeigii, Gaultheria forrestii, Hypericum forrestii, Jasminum polyanthum, Pieris formosa* var. *forrestii, Pyracantha rogersiana, Rhododendron griersonianum, R. sinogrande, R. forrestii, R. oreotrephes, Sorbus harrowiana, Syringa yunnanensis.*

Ernest 'Chinese' Wilson (1876–1930) Early expeditions in China and Japan for Veitch's nurseries, then for the Arnold Arboretum, United States between 1899 and 1919, finally from 1919 to 1921 in Australia, New Zealand, India and Africa. He discovered approximately 3,000 species of which more than 1,000 were new to cultivation, possibly the most famous being the handkerchief tree (*Davidia involucrata*) in China. Not forgetting *Lilium regale*, some of the loveliest garden shrubs and perennials like *Anemone hupehensis* var. *japonica.* Sponsored by Colonel Stephenson Clarke, Borde Hill received *Chamaecyparis formosensis* (planted in Warren Wood in 1912), *Acer fulvescens* (in North Park) and *Lindera megaphylla* (in South Roadside plantation). *Picea brachytyla* and *Pinus tabuliformis* were lost in the 1987 storm. He collected in North and South Korea, including two endemics, *Abies koreana* and *Rhododendron schlippenbachii.* Other plants amongst many: *Acer griseum, Berberis wilsoniae, Clematis montana* var. *rubens, Cornus kousa chinensis, Cotoneaster dammeri, C. salicifolius* var. *floccosus, Hydrangea sargentiana, Magnolia delavayi, Malus hupehensis, Lonicera nitida, Viburnum davidii, V.rhytidophyllum.*

Reginald Farrer (1880–1920) From 1914 to his death in 1920, he collected in China and Upper Burma (now Myanmar) with William Purdom and Euan Cox. His name is associated with *Gentiana farreri, Jasminum farreri, Juniperus recurva* var. *coxii*, *Viburnum farreri* and *Rosa farreri* aka Farrer's Threepenny Bit Rose (1915) propagated and offered by Knap Hill nurseries in 1934. Sponsored by Colonel Stephenson Clarke of Borde Hill.

Collingwood 'Cherry' Ingram (1880–1981) First visited Japan in 1902, returned five years later for his honeymoon with Florence. In 1919 they moved to The Grange, Benenden (in part of whose gardens is Balmoral Cottage, see pages 119–121) where he created a natural garden in which were two unusual mature flowering cherries. A passion was ignited, and he would introduce some 50 new varieties mostly from his own hybridisation, including the early flowering *Prunus* x *incam* 'Okame' and *P.* 'Kursar'. In 1925 a Japanese noble visited his garden and marvelled at one tree with large white flowers set against coppery leaves, he called it *tai-haku* or 'big white'. The next year Ingram returned to Japan, discovered that this *tai-haku* was lost to cultivation, and returned it to its rightful home. He raised *Rhododendron* and *Cistus* hybrids, and introduced new garden plants including *Rubus* 'Benenden' and rosemary (now *Salvia rosmarinifolia)* 'Benenden Blue' which he had collected in Corsica.

Frank Kingdon-Ward (1885–1958) Between 1911 and the 1940s he collected in Sichuan, Yunnan, Assam, Burma (Myanmar), Sikkim and Tibet. Apart from rhododendrons he collected *Meconopsis*, primulas, lilies and gentians. Sponsored by Colonel Stephenson Clarke, Borde Hill received many plants, notably the rhododendrons from his 1925 expedition planted in North Park. Other plants among many: *Acer wardii, Cornus chinensis, Cotoneaster wardii, Gaultheria wardii, Rhododendron wardii.*

***Davidia* handkerchiefs, Borde Hill**
Davidia involucrata in full handkerchief at Borde Hill where it springs from its original base. It was first discovered in 1869 by a French Jesuit missionary, Père David, whose name it celebrates, in Boaxing (Mupin) in China. 'Chinese' Wilson travelled 1,000 miles up the Yangtze river, further up the rapids above Ichang to Badong and then on foot following the tracks of salt-smugglers to find the tree had been felled. He retraced his steps collecting a few of the loveliest garden shrubs – *Azalea indica, Deutzia scabra, Jasminum floridum, Vitex negundo* and perennials like *Anemone hupehensis* var. *japonica*. He ventured south-west from Ichang into David's original territory where he 'drank in the beauties of this extraordinary tree' which 'stirred by the slightest breeze . . . [they] . . . resemble huge butterflies or small doves hovering'. (Courtesy Borde Hill)

THE HIGH WEALD COLLECTORS

Harold Frederick Comber (1897–1965) Born at Nymans, son of the head gardener, James Comber. Harold worked there after leaving school. Sir Edmund Loder of Leonardslee recommended him to Henry Elwes who engaged him at Colesbourne Park. Owing to a knee injury he was not called up during World War I. Loder and Elwes sponsored him to study at the Royal Botanic Garden Edinburgh. A worthwhile investment; he wrote a paper on the sterility of rhododendrons and gained 100 per cent in cryptogam botany. Between 1925 and 1927 he hunted twice in the Andes, collecting more than 1,200 species including the Chilean Fire Bush (*Embothrium coccineum*) and several species of *Berberis*, *Eucryphia* and *Fabiana*. In 1930 he collected seeds from 147 plants in Tasmania, sponsored by Colonel Stephenson Clarke of Borde Hill. In 1942 Clarke received an Award of Merit for his display of Comber's Tasmanian *Richea scoparia*. There is a contemporary connection, Stephen Herrington was the head gardener at Nymans until he took on this role at Leonardslee in September 2019. Herrington has retraced Comber's expeditions in Tasmania.

W.E.Th. Ingwersen Ltd In 1927 William Robinson provided land for Walter Ingwersen to set up a nursery which would become celebrated for its alpine plants. In the Borde Hill archives there is a flyer advertising '*Seeds from the Himalayas. Freshly collected this Autumn by Professor Arnold Wall, of Christchurch, New Zealand . . . 64 from Adonis chrysocyathus to Thalictrum cultratum*'. Stephenson Clarke ticked '*Cassiope fastigiata*; *Gentiana carinata from 12,000ft in the open 2s 6d; G. decumbens rather taller 1s 6d.*' In 1934 he thanks Stephenson Clarke for pointing out that gentians will not thrive in a limestone wall. In January 1936 he writes of an upcoming collecting trip to the '*Rockie mountains of British Columbia and N.W. Canada*' in June, July and August. His son Will Ingwersen (1905–1990) was a multi gold-medal winner at the Chelsea Flower Show and recognised authority on alpine and rockery plants.

***Cyclamen repandum*, Balmoral Cottage**
'Cherry' Ingram's sweetly fragrant Mediterranean *Cyclamen repandum* is colonising, offering delightful reddish pink flowers from March to May.

GENERAL INDEX

Page numbers in **bold** refer to illustrations

INDEX OF PLANTS

Page numbers in **bold** refer to illustrations

ACKNOWLEDGEMENTS

Writing this book has proved a garden of delights: renewing acquaintances, making new friends and enjoying warm welcomes in glorious settings. First and foremost, thanks to Andrewjohn and Eleni Stephenson Clarke with whom the idea of celebrating the English garden was born. They hosted me most generously enabling me to walk Borde Hill's pleasing gardens and landscape at all times of day, allowing me access to their archives and providing stunning photographs. Without exception all the garden owners warmly shared their enthusiasm and knowledge, pointed me in new and interesting directions and so I settled to writing with the intention of calling back for further photos. Then Covid-19 shut down 2020. The many images in this book that express how wildness pleases are thanks to these inspiring owners sharing their photographs, giving you the reader the chance to enjoy their gardens in many seasons. Special mentions, in order of their appearance in the book, to Lucy McLeod at Penshurst, Sarah Bray at High Beeches, Edward Barham at Hole Park, Nicola Talbot at Falconhurst, Hannah McDowell-Naylor at Wakehurst, Dan Luscombe at Bedgebury, Lisa Davies at Nymans, Tom Coward at Gravetye, Ingrid and Hans Sethi at Luctons, John and Linde Wotton at Goddards Green, Chris and Robin Hutt at Fairlight End, Mary Franck at Wych Warren, Fergus Garrett at Great Dixter, Imogen Stephens at Merriments, Patrick Rea and Stephen Herrington at Leonardslee, Clive and Jennie Whitmore at Twyford who also introduced me to the wonderful photography of Fiona McLeod and the food at The Cat in West Hoathly. I was sustained by excellent food at Gravetye Manor, Ashdown Park and Bedgebury Pinetum – all to be recommended.

ABOUT THE AUTHOR

Caroline Holmes is a garden historian of places, people and plants who likes to sift the humour from the humus but takes a pragmatic approach to numerology – this is her 13th book. She has researched and/or spoken on every continent except Antarctica. Her work is defined on three principles: to be academic but not dry, informative and entertaining, as well as practical and engaging. Her consultancy work specialises in exploring and evoking history. Over the years she has presented both BBC Radio and ITV Anglia television programmes from the High Weald as well as leading tours in French to key properties. Home is Suffolk, a one-acre garden and being surrounded by family and friends. www.horti-history.com

To Arthur
As he discovers the joy of gardening and
where the wildness pleases

ISBN 978-1-7888411-53

The author and publisher gratefully acknowledge the permission granted to reproduce the copyright material in this book. Every effort has been made to trace copyright holders and to obtain their permission for the use of copyright material. Any errors or omissions are entirely unintentional and should be addressed to the publisher.

British Library Cataloguing-in-Publication Data
A catalogue record for this book is available from the British Library

---

Front cover image: **Topiary, borders and house roofs, Great Dixter** (see page 116)

Back cover image: **Fairlight End garden** (see page 9). Courtesy Robin Hutt

Frontispiece: **Wheelbarrow path through meadow, Great Dixter**

Meadow flowers have to be left uncut in order to self-seed which can make getting a wheelbarrow across the garden something of a challenge, solved with beauty and utility in this delightful brick path at Great Dixter.

Page 4: **Lichen and moss on tree stumps, Wych Warren**

Healthy decay – a tree stump supports lichen and mosses with an artistic scattering of dark brown sweet chestnut and green oak leaves. Spent catkins and fungi add to the misty autumnal composition.

Page 175: **Wheelbarrow path, Twyford**

William Robinson favoured steps designed to match the garden areas as did Gertrude Jekyll, great if you are strolling around. This inset sloping brick is an ideal compromise for pushing the wheelbarrow around the garden or indeed cycling.

FSC
www.fsc.org
MIX
Paper from responsible sources
FSC® C106600

Printed in Slovenia
for ACC Art Books Ltd, Woodbridge, Suffolk, England
www.accartbooks.com

ACC ART BOOKS